On the path to AI

"Finding an analogy in the legal philosophy of Oliver Wendell Holmes Jr., the authors provide a penetrating and fine-grained examination of artificial intelligence, a rich and forward-looking approach that should restrain exaggerated claims and guide a realistic assessment of AI's prospects."

—Frederic R. Kellogg, *author of* Oliver Wendell Holmes Jr. and Legal Logic

"There's been a lot of discussion about how machine learning introduces or consolidates bias in AI due to its reliance on historic data. Who knew that law has been working on the social problems of the impact of precedent for over a century?"

—Joanna Bryson, *Professor of Ethics and Technology, Hertie School, Berlin*

Thomas D. Grant · Damon J. Wischik

On the path to AI

Law's prophecies and the conceptual foundations of
the machine learning age

Thomas D. Grant
Lauterpacht Centre for International
Law
University of Cambridge
Cambridge, UK

Damon J. Wischik
Department of Computer Science and
Technology
University of Cambridge
Cambridge, UK

ISBN 978-3-030-43581-3   ISBN 978-3-030-43582-0   (eBook)
https://doi.org/10.1007/978-3-030-43582-0

This Palgrave Macmillan imprint is published by the registered company Springer Nature
Switzerland AG
The registered company address is: Gewerbestrasse 11, 6330 Cham, Switzerland

Prologue—Starting with Logic

Law and computer science, in their classic form, employ logic to produce results. Edsger Dijkstra (1930–2002), one of the pioneers of computer science, expressed the essence of the field in its earlier times. Computer science had at its heart the mathematical analysis of algorithms, and thus...

> Programming is one of the most difficult branches of applied mathematics; the poorer mathematicians had better remain pure mathematicians.[1]

This pithy summation of computer science—folding it into a branch of applied mathematics—is resilient. It remains at the root of the widespread view of computer science, still taught in undergraduate courses and echoed in explanations to the general public, that *computers run algorithms, which are step-by-step instructions for performing a task*.[2] They might be complicated, too hard for "poorer mathematicians" to understand, but in the end they are formulaic. They are logical processes, readily designed, readily evaluated for success or failure, and readily fixed, so long as you have the analytic skills needed to understand their logic.

In law, thinking long followed lines much like those Dijkstra described in computer science. Not least among the early modern exponents of classic legal thinking, there was Sir William Blackstone (1723–1780), who, when writing *On the Study of the Law*, set out as belonging to the essentials that the student...

can reason with precision, and separate argument from fallacy, by the clear simple rules of pure unsophisticated logic ... can fix his attention, and steadily pursue truth through any the most intricate deduction, by the use of mathematical demonstrations ... [and] has contemplated those maxims reduced to a practical system in the laws of imperial Rome ...[3]

The well-schooled lawyer, like the better mathematician, gets to the correct result as surely as the well-designed algorithm generates a satisfactory computer output. As intricate as the deductions might be and thus demanding on the intellect, the underlying process is "pure unsophisticated logic."

But to sum up computer science that way is out of date. The current boom in artificial intelligence, driven by machine learning, is not about deducing logical results from formulae. It is instead based on inductive prediction from datasets. The success of machine learning does not derive from better mathematics. It derives instead from bigger datasets and better understanding of the patterns those datasets contain. In the chapters that follow, we will explore this revolution in computer science.

A revolution has taken place in modern times in law as well. As one would expect of a shift in thinking which has far-reaching impact, more than one thinker has been involved. Nevertheless, in law, one figure over the past century and a half stands out. Oliver Wendell Holmes Jr., the title of whose famous essay *The Path of the Law* we borrow in paraphrase for the title of our book, influenced the law and motivated changes in how people think about the law. To such an extent did Holmes affect legal thinking that his work marks a turning point.

We believe that the revolution in computer science that machine learning entails mirrors the revolution in law in which Oliver Wendell Holmes Jr. played so prominent a part. This book describes both revolutions and draws an analogy between them. Our purpose here is to expose the fundamental contours of thought of the machine learning age—its conceptual foundations—by showing how these trace a similar shape to modern legal thought; and by placing both in their wider intellectual setting. Getting past a purely technological presentation, we will suggest that machine learning, for all its novelty and impact, belongs to a long history of change in the methods people use to make sense of the world. Machine learning is a revolution in thinking. It has not happened, however, in isolation.

Machine learning deserves an account that relates it both to its immediate antecedents in computer science and to another socially vital endeavor. Society at large deserves an account that explains what machine learning really is.

HOLMES AND HIS LEGACY

As the nineteenth century drew to a close in America, growth and change characterized practically every field of endeavor. Legal education partook of the upward trend, and the Boston University School of Law, then still a relative newcomer in the American city that led the country in academic endeavor, built a new hall. To mark the opening of the new hall, which was at 11 Ashburton Place, the dean and overseers of the School invited Holmes to speak. Then aged 55 and an Associate Justice of the Massachusetts Supreme Judicial Court, Holmes was a local luminary, and he could be counted on to give a good speech. There is no evidence that the School was looking for more than that. The speech that Holmes gave on January 8, 1897, however, pronounced a revolution in legal thought. Its title was *The Path of the Law*.[4] Published afterward in the Harvard Law Review, this went on to become one of the most cited works of any jurist.[5] *The Path of the Law*, not least of all Holmes's statement therein that the law is the "prophecies of what the courts will do in fact," exercises an enduring hold on legal imagination.[6] Holmes rejected "the notion that a [legal system]… can be worked out like mathematics from some general axioms of conduct."[7] He instead defined law as consisting of predictions or "prophecies" found in the patterns of experience. From its starting point as an operation of logical deduction, law according to Holmes, if law were to be understood fully, had to be understood as something else. It had to be understood as a process of induction with its grounding in modern ideas of probability.

Holmes's earlier postulate, that "[t]he life of the law has not been logic; it has been experience,"[8] likewise has been well-remembered.[9] Holmes was not telling lawyers to make illogical submissions in court or to give their clients irrational advice. Instead, he meant to lead his audience to new ways of thinking about their discipline. Law, in Holmes's view, starts to be sure from classic logic, but logic gets you only so far if you hope to understand the law.

Holmes lived from 1841 to 1935, and so longevity perhaps contributed to his stature. There was also volume of output. Holmes authored over

800 judgments, gave frequent public addresses many of which are set down in print, and was a prolific correspondent with friends, colleagues, and the occasional stranger.[10] There is also quotability.[11] Holmes has detractors[12] and champions.[13] He has been the subject of "cycles of intellectual anachronisms, panegyrics, and condemnations."[14] It is not to our purpose to add to the panegyrics or to the condemnations. We do wholeheartedly embrace anachronism! Actually, we do not deny the limits of analogy across two disciplines across over a century of change; we will touch on some of the limits (Chapter 3). Yet, even so, Holmes's conception of the law, in its great shift from formal deduction to inductive processes of pattern searching, prefigured the change from traditional algorithmic computing to the machine learning revolution of recent years. And, going further still, Holmes posited certain ideas about the process of legal decision making—in particular about the effect of past decisions and anticipated future decisions on making a decision in a case at hand—that suggest some of the most forward-thinking ideas about machine learning that computer scientists are just starting to explore (Chapter 9). There is also a line in Holmes's thought that queried whether, notwithstanding the departure from formal proof, law might someday, through scientific advances that uncover new rules, find its way back to its starting point in logic. Here too an inquiry that Holmes led over a century ago in law may be applied today as we consider what the machine learning age holds in store (Chapter 10).

As for the law in his day as Holmes saw it, and as many have since, it must be seen past its starting point in deductive reasoning if one is to make sense of it.[15] Law is, according to Holmes, not logic, but *experience*— meaning that the full range of past decisions, rules, and social influences is what really matters in law. An "inductive turn" in Holmes's thinking about law[16]—and in law as practiced and studied more widely—followed. In computer science, the distinct new factor has been the emergence of *data* as the motive force behind machine learning. How much weight is to be attributed to logic, and how much to experience or data, is a point of difference among practitioners both in law and in computer science. The difference runs deep in the history and current practice of the fields, so much so that in law and in computer science alike it marks a divide in basic understandings.

Jurists refer to formalists and realists when describing the divide in legal understanding that concerns us here. The formalists understand law as the application of logical rules to particular questions. The realists see it, instead, as the discovery of patterns of behavior in a variety of legal and social sources. The formalists see their approach to law as the right place to start, and the strictest among them see the emergence of legal realism as a setback, not an advance, for law. The realists, for their part, sometimes dismiss the formalists as atavistic. The divide runs through the professional communities of advocates, advisers, and judges as much as through legal academia.[17]

The divide in computer science is neither as storied nor as sharply defined as that in law. It is not associated with any such widely accepted monikers as those attached to the logic-based formalists or the pattern-seeking realists in law. As we will explore further below, only in recent years has computing come to be a data-driven process of pattern finding. Yet the distinction between the deductive approach that is the basis of classic computer algorithms, and the inductive approach that is the basis of present-day advances in machine learning, is the central distinction in what may prove to be the central field of technological endeavor of the twenty-first century. The emergence of machine learning will be at best imperfectly understood if one does not recognize this conceptual shift that has taken place.

What is involved here is no less than two great revolutions in theory and practice, underway in two seemingly disparate fields but consisting in much the same shift in basic conception. From conceiving of law and computer science purely as logical and algorithmic, people in both fields have shifted toward looking for patterns in experience or data. To arrive at outcomes in either field still requires logic but, in the machine learning age, just as in the realist conception of law that emerged with Holmes, the path has come to traverse very different terrain.

A Note on Terminology: Machine Learning, Artificial Intelligence, and Neural Networks

In this book, we will refer to *machine learning*. Our goal in the following chapters is to explain what machine learning is—but before proceeding it may be helpful to say a few words to clarify the difference between machine learning, artificial intelligence, and neural networks.[18]

Artificial intelligence refers in academia to an evolving field which encompasses many areas from symbolic reasoning to neural networks. In popular culture, it encompasses everything from classic statistics re-branded by a marketing department ("Three times the AIs as the next leading brand!") to science fiction, invoking pictures of robotic brains and conundrums about the nature of intelligence.

Machine learning is a narrower term. The United Kingdom House of Lords Select Committee on Artificial Intelligence in its 2018 report[19] highlights the difference: "The terms 'machine learning' and 'artificial intelligence' are ... sometimes conflated or confused, but machine learning is in fact a particular type of artificial intelligence which is especially dominant within the field today." The report goes on to say, "We are aware that many computer scientists today prefer to use 'machine learning' given its greater precision and lesser tendency to evoke misleading public perceptions." Broadly speaking, machine learning is the study of computer systems that use systematic mathematical procedures to find patterns in large datasets and that apply those patterns to make predictions about new situations. Many tools from classical statistics can be considered to be machine learning, though machine learning as an academic discipline can be said to date from the 1980s.[20]

Artificial neural network refers to a specific design of machine learning system, loosely inspired by the connections of neurons in the brain. The first such network, the Perceptron, was proposed by Frank Rosenblatt of the Cornell Aeronautical Laboratory in 1958.[21] The original Perceptron had a simple pattern of connections between its neurons. Networks with more complex patterns are called *deep neural networks*, and the mathematical procedure by which they learn is called *deep learning*. The current boom[22] in artificial intelligence is based almost entirely on deep learning, and one can trace it to a single event: in 2012, in an annual competition called the ImageNet Challenge, in which the object is to build a computer program to classify images,[23] a deep neural network called AlexNet[24] beats all the other competitors by a significant margin. Since then, a whole host of tasks, from machine translation to playing Go, have been successfully tackled using neural networks. It is truly remarkable that these problems can be solved with machine learning, rather than requiring some grander human-like generalartificial intelligence. The reason it took from 1958 to 2012 to achieve this success is mostly attributable to computer hardware limitations: it takes a huge amount of processing on big datasets for deep learning to work, and it was only in 2012 that computer hardware's exponential improvement met the needs of image classification.[25] It also has helped that the means for gathering and storing big datasets have improved significantly since the early days.

In this book, we will use the term *machine learning*, and we will not stray any further into artificial intelligence. We have neural networks in mind, but our discussion applies to machine learning more widely. Whether machine learning and neural networks have a role to play in the possible future emergence of a *general* AI—that is to say, a machine that approximates or exceeds human intelligence—we will not even speculate.[26]

NOTES

1. Dijkstra, *How Do We Tell Truths That Might Hurt?* in Dijkstra, SELECTED WRITINGS ON COMPUTING: A PERSONAL PERSPECTIVE (1982) 129 (original text dated June 18, 1975).

2. Countless iterations of this description appear in course materials on computer programming. See, e.g., http://computerscience.chemeketa.edu/cs160Reader/Algorithms/AlgorithmsIntro.html; http://math.hws.edu/javanotes/c3/s2.html. For a textbook example, see Schneider & Gersting, INVITATION TO COMPUTER SCIENCE (1995) 9. Schneider and Gersting, in their definition, stipulate that an algorithm is a "well-ordered collection of unambiguous and effectively computable operations that when executed produces a result and halts in a finite amount of time." We will say some more in Chapter 9 about the "halting problem": see p. 109.

3. Sir William Blackstone, COMMENTARIES ON THE LAWS OF ENGLAND: BOOK THE FIRST (1765) 33.

4. Holmes, *The Path of the Law,* 10 Harv. L. Rev. 457 (1896–97).

5. Fred R. Shapiro in *The Most-Cited Law Review Articles Revisited,* 71 CHI.-KENT L. REV. 751, 767 (1996) acknowledged that the fifth-place ranking of *The Path of the Law* reflected serious undercounting, because the only citations counted were those from 1956 onward. Only a small handful of Shapiro's top 100 were published before 1956. Shapiro and his co-author Michelle Pearse acknowledged a similar limitation in a later update of the top citation list: *The Most-Cited Law Review Articles of All Time,* 110 MICH. L. REV. 1483, 1488 (2012). *The Path* came in third in Shapiro and Pearse's 2012 ranking: id. at 1489.

6. Illustrated, for example, by the several symposiums on the occasion of its centennial: See 110 HARV. L. REV 989 (1997); 63 BROOK. L. REV. 1 (1997); 78 B.U. L. Rev 691 (1998) (and articles that follow in each volume). Cf. Alschuler, 49 FLA. L. REV. 353 (1997) (and several responses that follow in that volume).

7. Holmes, *The Path of the Law,* 10 HARV. L. REV. 457, 465 (1896–97).

8. Holmes, THE COMMON LAW (1881) 1.

9. A search discloses over three hundred instances of American judges quoting the phrase in judgments, some three dozen of these being judgments of the U.S. Court of Appeals, some half dozen of the U.S. Supreme Court.

10. Thus, the 1995 collection edited by Novick of mostly non-judicial writings (but not personal correspondence) runs to five volumes.

11. It would fill over a page to give citations to American court judgments, state and federal, referring to Holmes as "pithy" (from Premier-Pabst Sales Co. v. State Bd. of Equalization, 13 F.Supp. 90, 95 (District Court, S.D. California, Central Div.) (Yankwich, DJ, 1935) to United States v. Thompson, 141 F.Supp.3d 188, 199 (Glasser, SDJ, 2015)) or "memorable" (from Regan & Company, Inc. v. United States, 290 F.Supp.470, (District Court, E.D. New York) (Rosling, DJ, 1968) to Great Hill Equity Partners IV, et al. v. SIG Growth Equity Fund I, et al., (unreported, Court of Chancery, Delaware) (Glasscock, VC, 2018)). On aesthetics and style in Holmes's writing, see Mendenhall, *Dissent as a Site of Aesthetic Adaptation in the Work of Oliver Wendell Holmes Jr.*, 1 BRIT. J. AM. LEGAL STUD. 517 (2012) esp. id. at 540–41.

12. For example Ronald Dworkin & Lon L. Fuller. See Ronald Dworkin, LAW'S EMPIRE (1986) 13–14; Lon Fuller, *Positivism and Fidelity to Law—A Reply to Professor Hart,* 71 HARV. L. REV. 630 esp. id. at 657–58 (1958).

13. Richard A. Posner is perhaps the most prominent of the champions in the late twentieth and early twenty-first centuries. See Posner's *Introduction* in THE ESSENTIAL HOLMES. SELECTIONS FROM THE LETTERS, SPEECHES, JUDICIAL OPINIONS, AND OTHER WRITINGS OF OLIVER WENDELL HOLMES, JR. (1992). Cf. H.L.A. Hart, *Positivism and the Separation of Law and Morals,* 71 HARV. L. REV. 593 (1958) (originally the Oliver Wendell Holmes Lecture, Harvard Law School, April 1957). Further to a curious link that Hart seems to have supplied between Holmes and computer science, see Chapter 10, p. 123.

14. Pohlman (1984) 1. Cf. Gordon (1992) 5: Holmes has "inspired, and… continues to inspire, both lawyers and intellectuals to passionate attempts to come to terms with that legend—to appropriate it to their own purposes, to denounce and resist it, or simply to take it apart to see what it is made of."

15. Holmes, THE COMMON LAW (1881) 1.

16. The apt phrase "inductive turn" is the one used in the best treatment of Holmes's logic: Frederic R. Kellogg, OLIVER WENDELL HOLMES JR. AND LEGAL LOGIC (2018) pp. 35, 72–87, about which see further Chapter 1, p. 2.

17. For a flavor of the critique of formalism, see Frederick Schauer's treatment of the Supreme Court's judgment in *Lochner v. New York* and its reception: Frederick Schauer, *Formalism,* 97 YALE L. J. 509, 511–14 (1988); and

for frontal defenses (albeit from very different quarters), Antonin Scalia, *The Rule of Law as a Law of Rules*, 56 U. CHI. L. REV. 1175 (1989) and James Crawford, *Chance, Order, Change: The Course of International Law*, in Hague Academy of Int'l Law, 365 RECUEIL DES COURS 113, 113–35 (2013). Further to formalism, see Chapter 1, p. 2; Chapter 2, pp. 20–21.

18. The 2018 *Report of the UN Secretary-General on Current developments in science and technology and their potential impact on international security and disarmament efforts* put the relation between the terms like this:

> Modern artificial intelligence comprises a set of sub-disciplines and methods that leverage technology, such as data analysis, visual, speech and text recognition, and robotics. Machine learning is one such sub-discipline. Whereas hand-coded software programmes typically contain specific instructions on how to complete a task, machine learning allows a computer system to recognize patterns in large data sets and make predictions. Deep learning a subset of machine learning, implements various machine-learning techniques in layers based on neural networks, a computational paradigm loosely inspired by biological neurons. Machine-learning techniques are highly dependent on the quality of their input data, and arguably the quality of the data is more important to the success of a system than is the quality of the algorithm. A/73/177 (July 17, 2018). Cf. Chapter 1, p. 14, n. 12.

The proper distinction between the three terms has led to heated argument between technologists. See e.g., https://news.ycombinator.com/item?id=20706174 (accessed Aug. 24, 2019).

19. Select Committee on Artificial Intelligence (Lords), Report (Apr. 16, 2018) p. 15, 17.
20. See, e.g., Efron & Hastie (2016) 351. See also Leo Breiman as quoted in Chapter 1, p. 1.
21. Rosenblatt (1958).
22. Artificial intelligence has experienced a series of booms and "AI winters." For a broader history of artificial intelligence, see Russell & Norvig (2016) 5–27. Cf. National Science and Technology Council (U.S.), *The National Artificial Intelligence Research Development Strategic Plan* (Oct. 2016) pp. 12–14, describing three "waves" of AI development since the 1980s.
23. The ImageNet database was announced in 2009: J. Deng, W. Dong, R. Socher, L.-J. Li, K. Li & L. Fei-Fei, ImageNet: A Large-Scale Hierarchical Image Database, CVPR, 2009. The first ImageNet Challenge was in 2010. For the history of the Challenge, see Olga Russakovsky*, Jia Deng*, Hao Su, Jonathan Krause, Sanjeev Satheesh, Sean Ma, Zhiheng Huang, Andrej

Karpathy, Aditya Khosla, Michael Bernstein, Alexander C. Berg and Li Fei-Fei. (* = equal contribution) ImageNet Large Scale Visual Recognition Challenge. IJCV, 2015.

24. Krizhevksy, Sutskever & Hinton (2017) 60(6) Comms. Acm 84–90.

25. The computational power comes from better hardware in the form of graphics processing units (GPUs). The computer gaming industry spurred the development of hardware for better graphics, and this hardware was then used to speed up the training of neural networks.

   Cognitive neuroscientists have observed a correlation between the development of eyes and brain size: See, e.g., Gross, *Binocularity and Brain Evolution in Primates*, (2004) 101(27) pnas 10113–15. See also Passingham & Wise, THE NEUROBIOLOGY OF THE PREFRONTAL CORTEX: ANATOMY, EVOLUTION AND THE ORIGIN OF INSIGHT (2012). Thus in some, albeit very general sense, a link is suggested both in biological evolution and in the development of computer science between the increase in processing power (if one permits such an expression in regard to brains as well as GPUs) and the demands of dealing with imagery.

26. For speculation about an impending "singularity"—a future moment when AI emerges with capacities exceeding human cognition—see Bostrom, SUPERINTELLIGENCE: PATHS, DANGERS, STRATEGIES (2014).

CONTENTS

About the Authors

Thomas D. Grant is a Fellow of the Lauterpacht Centre for International Law, University of Cambridge.

Damon J. Wischik is a Lecturer in the Department of Computer Science and Technology, University of Cambridge, and a Fellow of the Alan Turing Institute, London.

ABBREVIATIONS

ACM	Association for Computing Machinery
ACM CSUR	ACM Computing Surveys
AJIL	American Journal of International Law
Art. Intel. & Law	Artificial Intelligence and Law (Journal)
B.U. L. Rev.	Boston University Law Review
Brook. L. Rev.	Brooklyn Law Review
Cal. L. Rev.	California Law Review
Chi.-Kent L. Rev.	Chicago Kent Law Review
Col. L. Rev.	Columbia Law Review
Comms. ACM	Communications of the ACM
Corn. Int'l L.J.	Cornell International Law Journal
Corn. L. Rev.	Cornell Law Review
Crim. L. Rev.	Criminal Law Review
CVPR	Computer Vision and Pattern Recognition
EJIL	European Journal of International Law
Fla. L. Rev.	Florida Law Review
GDPR	General Data Protection Regulation (European Union)
Geo. L.J.	Georgetown Law Journal
Geo. Wash. L. Rev.	George Washington Law Review
GPUs	Graphics processing units
GYIL	German Yearbook of International Law
Harv. J.L. & Tech.	Harvard Journal of Law and Technology
Harv. L. Rev. F.	Harvard Law Review Forum
Harv. L. Rev.	Harvard Law Review

ICSID	International Centre for the Settlement of Investment Disputes
IEEE	Institute of Electrical and Electronics Engineers
IMA Bull.	Bulletin of the IMA
IMA	Institute of Mathematics and Its Applications
ITCS Conf. Proc. (3rd)	Proceedings of the 3rd ITCS Conference
ITCS	Innovations in Theoretical Computer Science
J. Evol. & Tech.	Journal of Evolution and Technology
J. Pat. & Trademark Off. Soc'y	Journal of Patent and Trademark Office Society
LMS Proc.	Proceedings of the London Mathematical Society
Md. L. Rev.	Maryland Law Review
Mich. L. Rev.	Michigan Law Review
MLR	Modern Law Review
N.D. L. Rev.	North Dakota Law Review
N.Y.U.L. Rev.	New York University Law Review
NEJM	New England Journal of Medicine
Nw. U. L. Rev.	Northwestern University Law Review
Phil. Trans. R. Soc. A	Philosophical Transactions of the Royal Society A: Mathematical, Physical and Engineering Science
PNAS	Proceedings of the National Academy of Sciences
Stan. L. Rev.	Stanford Law Review
Sw. J. Int'l L.	Southwestern Journal of International Law
Temp. L.Q.	Temple Law Quarterly
Tex. L. Rev.	Texas Law Review
U. Chi. L. Rev.	University of Chicago Law Review
U. Pa. L. Rev.	University of Pennsylvania Law Review
U. Pitt. L. Rev.	University of Pittsburgh Law Review
UP	University Press
Vill. L. Rev.	Villanova Law Review
Yale L.J.	Yale Law Journal

CHAPTER 1

Two Revolutions

What constitutes the law? You will find some text writers telling you that it is something different from what is decided by the courts of Massachusetts or England, that it is a system of reason, that it is a deduction from principles of ethics or admitted axioms or what not, which may or may not coincide with the decisions. But if we take the view of our friend the bad man we shall find that he does not care two straws for the axioms or deductions, but that he does want to know what the Massachusetts or English courts are likely to do in fact. I am much of his mind. The prophecies of what the courts will do in fact, and nothing more pretentious, are what I mean by the law.

Oliver Wendell Holmes, Jr. The Path of the Law *(1897)*

In the mid-1980s two powerful new algorithms for fitting data became available: neural nets and decision trees. A new research community using these tools sprang up. Their goal was predictive accuracy. The community consisted of young computer scientists, physicists and engineers plus a few aging statisticians. They began using the new tools in working on complex prediction problems where it was obvious that data models were not applicable: speech recognition, image recognition, nonlinear time series prediction, handwriting recognition, prediction in financial markets.

Leo Breiman, Statistical Modeling: The Two Cultures *(2001)*

Machine learning, the method behind the current revolution in artificial intelligence,[1] may serve a vast range of purposes. People across practically every walk of life will feel its impact in the years to come. Not many of them, however, have any very clear idea how machine learning works.

© The Author(s) 2020
T. D. Grant and D. J. Wischik, *On the path to AI*,
https://doi.org/10.1007/978-3-030-43582-0_1

The present short book describes how machine learning works. It does so with a surprising analogy.

Oliver Wendell Holmes, Jr., one of the law's most influential figures in modern times, by turns has been embraced for the aphoristic quality of his writing and indicted on the charge that he reconciled himself too readily to the injustices of his day. It would be a mistake, however, to take Holmes to have been no more than a crafter of *beaux mots*, or to look no further than the judgment of some that he lacked moral compass. That would elide Holmes's role in a revolution in legal thought—and the remarkable salience of his ideas for a revolution in computer science now under way.

Holmes in the years immediately after the American Civil War engaged with leading thinkers of the nineteenth century, intellectuals who were taking a fresh look at scientific reasoning and logic and whose insights would influence a range of disciplines in the century to come. The engagement left an imprint on Holmes and, through his work as scholar and as judge, would go on to shape a new outlook on law. Holmes played a central role in what has recently been referred to as an "inductive turn" in law,[2] premised on an understanding that law in practice is not a system of syllogism or formal proof but, instead, a process of discerning patterns in experience. Under his influence, legal theory underwent a change from deduction to induction, from formalism to realism. This change has affected the law in theory and in practice. It is oft-recounted in modern legal writing. A formalist view of legal texts—seeing the law as a formula that can be applied to the factual situations the legislator promulgated the law to address—remains indispensable to understanding law; but formalism, for better or worse, no longer suffices if one is to understand how lawyers, judges, and others involved in the law actually operate.

In computer science, a change has occurred which today is having at least as much impact, but in most quarters remains unknown or, at best, imprecisely grasped. The new approach is an inductive data-driven approach, in which computers are "trained" to make predictions. It had its roots in the 1950s and has come to preeminence since 2012. The classic view of computing, by contrast, is that computers execute a series of logical steps that, applied to a given situation, lead to completion of a required task; it is the programmer's job to compose the steps as an algorithm to perform the required task. Machine learning is still built on computers that execute code as a series of logical steps, in the way they have since the start of modern computing[3]—but this is not an adequate explanation of what makes machine learning such a powerful tool, so powerful that people talk of it as the point of departure toward a genuine artificial intelligence.

1.1   An Analogy and Why We're Making It

In this book, we describe, in the broadest sense, how machine learning does what it does. We argue that the new and unfamiliar terrain of machine learning mirrors with remarkable proximity Holmes's conception of the law. Just as the law is a system of "prophesy from experience," as Holmes put it, so too machine learning is an inductive process of prediction based on data. We consider the two side by side in the chapters that follow for two mutually supporting purposes: in order to convey a better understanding of machine learning; and in order to show that the concepts behind machine learning are not a sudden arrival but, instead, belong to an intellectual tradition whose antecedents stretch back across disciplines and generations.

We will describe how machine learning differs from traditional algorithmic programming—and how the difference between the two is strikingly similar to the difference between the inductive, experience-based approach to law so memorably articulated by Holmes and the formalist, text-based approach that that jurist contrasted against his own. Law and computing thus inform one another in the frame of two revolutions in thought and method. We'll suggest why the likeness between these two revolutions is not happenstance. The changes we are addressing have a shared origin in the modern emergence of ideas about probability and statistics.

Those ideas should concern people today because they have practical impact. Lawyers have been concerned with the impact of the revolution in their own field since Holmes's time. It is not clear whether technologists' concern has caught up with the changes machine learning has brought about. Technologists should concern themselves with machine learning not just as a technical project but also as a revolution in how we try and make sense of the world, because, if they don't, then the people best situated to understand the technology won't be thinking as much as they might about its wider implications. Meanwhile, social, economic, and political actors need to be thinking more roundly about machine learning as well. These are the people who call upon our institutions and rules to adapt to machine learning; some of the adaptations proposed to date are not particularly well-conceived.[4]

New technologies of course have challenged society before the machine learning age. Holmes himself was curious and enthusiastic about the technological change which over a hundred years ago was already unsettling so many of the expectations which long had lent stability to human relations. He did not seem worried about the downsides of the innovations that roused his interest. We turn to Holmes the futurist in the concluding part of this book by way of postscript[5]; scientism—an unexamined belief that science and technology can solve any problem—is not new to the present era of tech-utopians.

Our principal concern, however, is to foster a better understanding of machine learning and to locate this revolution in its wider setting through an analogy with an antecedent revolution in law. We address machine learning because those who make decisions about this technology, whether they are concerned with its philosophical implications, its practical potential, or safeguards to mitigate its risks, need to know what they are making decisions about. We address it the way we do because knowledge of a thing grows when one sees how it connects to other things in the world around it.

1.2   What the Analogy Between a Nineteenth Century Jurist and Machine Learning Can Tell Us

The claim with which we start, which we base on our understanding of the two fields that the rest of this book will consider, is this: Holmes's conception of the law, which has influenced legal thought for close to a century and a half, bears a similar conceptual shape and structure to that which computing has acquired with the recent advances in machine learning. One purpose in making this claim is to posit an analogy between a change in how people think about law, and a change that people need to embrace in their thinking about how computers work—if they are to understand how computers work in the present machine learning age. The parallels between these two areas as they underwent profound transformation provide the organizing idea of this book.

Despite the myriad uses for machine learning and considerable attention it receives, few people outside immediate specialty branches of computer science and statistics avoid basic misconceptions about what it is. Even within the specialties, few experts have perspective on the conceptual re-direction computer science in recent years has taken, much less an awareness of its kinship to revolutionary changes that have shaped another socially vital field. The analogy that we develop here between law and machine learning supplies a new way of looking at the latter. In so doing, it helps explain what machine learning is. It also helps explain where machine learning comes from: the recent advances in machine learning have roots that reach deeply across modern thought. Identifying those roots is the first step toward an intellectual history of machine learning. It is also vital to understanding why machine learning is having such impact and why it is likely to have still more in the years ahead.

The impact of machine learning, realized and anticipated, identifies it as a phenomenon that requires a social response. The response is by no means limited to law and legal institutions, but arriving at a legal classification of the phenomenon is overdue. Lawyers and judges already are called upon to address machine learning with rules.[6] And, yet, legislative and regulatory authorities are at a loss for satisfactory definition. We believe that an analogy between machine learning and law will help.

But what does an analogy tell us, that a direct explanation does not?

One way to gain understanding of what machine learning *is* is by enumerating what it *does*. Here, for example, is a list of application areas supplied by a website aimed at people considering careers in data science:

- Game-playing
- Transportation (automated vehicles)
- Augmenting human physical and mental capabilities ("cyborg" technology)
- Controlling robots so they can perform dangerous jobs
- Protecting the environment
- Emulating human emotions for the purpose of providing convincing robot companions
- Improving care for the elderly
- General health care applications
- Banking and financial services
- Personalized digital media
- Security
- Logistics and distribution (supply chain management)
- Digital personal assistants
- E-commerce
- Customizing news and market reports.[7]

Policy makers and politicians grappling with how to regulate and promote AI make lists like this too. The UK House of Lords, for example, having set up a Select Committee on Artificial Intelligence in 2017, published a report of the Committee which, *inter alia*, listed a number of specific fields which are using AI.[8] An Executive Order of the President of the United States, adopted in 2019, highlighted the application of AI across diverse aspects of the national economy.[9] The People's Republic of China Ministry of Industry and Information Technology adopted an Action Plan in 2017 for AI which identified a range of specific domains in which AI's applications are expected to grow.[10] But while lists of applications can reflect where the technology is used today, they don't indicate where it might or might not be used in the future. Nor do such lists convey the clearer understanding of how AI works that we need if we are to address it, whether our purpose is to locate AI in the wider course of human development to which it belongs or to adjust our institutions and laws so that they are prepared for its impact, purposes which, we suggest, are intertwined. AI is a tool, and naming things the tool does is at best only a roundabout route to defining it.

Suggesting the limits in that approach, others attempting to define artificial intelligence have not resorted to enumeration. To give a high profile example, the European Commission, in its Communication in 2018 on *Artificial Intelligence for Europe*, defined AI as "systems that display intelligent behavior by analyzing their environment and taking actions—with some degree of autonomy—to achieve specific goals."[11] This definition refers to AI as technology "to achieve specific goals"; it does not list what those goals might be. It is thus a definition that places weight not on applications (offering none of these) but instead on general characteristics of what it defines. However, defining AI as "systems that display intelligent behaviour" is not adequate either; it is circular. Attempts to define machine learning and artificial intelligence tend to rely on synonyms that add little to a layperson's understanding of the computing process involved.[12] In a machine learning age, more is needed if one is both to grasp the technical concept and to intuit its form.

In this book, we do not continue the search for synonyms or compile an index of extant definitions. Nor do we undertake to study how AI might be applied to particular practical problems in law or other disciplines. Instead, we aim to develop and explore an analogy that will help people understand machine learning.

The value of analogy as a means to understand this topic is suggested when one considers how definitions of unfamiliar concepts work. Carl Hempel, one of the leading thinkers on the philosophy of science in the twentieth century, is known by American lawyers for the definition of "science" that the U.S. Supreme Court espoused in *Daubert*, a landmark in twentieth century American jurisprudence.[13] Hempel was concerned as well with the definition of "definition." He argued that definition *"requires the establishment of diverse connections... between different aspects of the empirical world."*[14] It is from the idea of diverse connections that we take inspiration. We posit an analogy between two seemingly unrelated fields and with that analogy elucidate the salient characteristics of an emerging technology that is likely to have significant effects on many fields in the years to come.[15] We aim with this short book to add to, and diversify, the connections among lawyers, computer scientists, and others as well, who should be thinking about *how* to think about the machine learning age which has now begun.

We will touch on the consequences of the change in shape of both law and computing, but our main concern lies elsewhere—namely, to supply the reader with an understanding of how precisely under a shared intellectual influence those fields changed shape and, moreover, with an understanding of what machine learning—the newer and less familiar field—is.

1.3   APPLICATIONS OF MACHINE LEARNING IN LAW—AND EVERYWHERE ELSE

Writers in the early years of so-called artificial intelligence, before machine learning began to realize its greater potential, were interested in how computers might affect legal practice.[16] Many of them were attempting to find ways to use AI to perform particular law-related tasks. Some noted the formalist-realist divide that had entered modern legal thinking.[17] Scholars and practitioners who considered law and AI were interested in the contours of the former because they wished to see how one might get a grip on it using the latter, like a farmer contemplating a stone that she needs to move and reckoning its irregularities, weight, position, etc. before hooking it up to straps and pulleys. Thus, to the extent they were interested in the nature of law, it was because they were interested in law as a possible object to which to apply AI, not as a source of insight into the emergence of machine learning as a distinct way that computers might be used.

Investigation into practical applications of AI, including in law, has been reinvigorated by advances that machine learning has undergone in recent years. The advances here have taken place largely since 2012.[18] In the past several years, it seems scarcely a day goes by without somebody suggesting that artificial intelligence might supplement, or even replace, people in functions that lawyers, juries, and judges have performed for centuries.[19] In regard to functions which the new technology already widely performs, it is asked what "big data" and artificial intelligence imply for privacy, discrimination, due process, and other areas of concern to the law. An expanding literature addresses the tasks for which legal institutions and the people who constitute them use AI or might come to in the future, as well as strategies that software engineers use, or might in the future, to bring AI to bear on such tasks.[20] In other words, a lot is being written today about AI and law *as such*. The application of AI in law, to be sure, has provoked intellectual ferment, certain practical changes, and speculation as to what further changes might come.

But the need for a well-informed perspective on machine learning is not restricted to law. We do not propose here to address, much less to solve, the technical challenges of putting AI in harness to particular problems, law-related or other. It is not our aim here to compile another list of examples of tasks that AI performs, any more than it is our purpose to list examples of the subject matter that laws regulate. Tech blogs and policy documents, like the ones we just referred to above, are full of suggestions as to the former; statute books and administrative codes contain the latter. Nor is it our purpose here to come up with programming strategies for the application of AI to tasks in particular fields; tech entrepreneurs and software engineers are doing that in law and many fields besides.

There are law-related problems—and others—that people seek to employ machine learning to solve, but cataloguing the problems does not in itself impart much understanding of what machine learning *is*. The concepts that we deal with here concern *how* the mechanisms work, not (or not primarily) what they might do (or what problems they might be involved in) when they work. Getting at these concepts is necessary, if the people who ought to understand AI today, lawyers included, are actually to understand it. Reaching an understanding of how its mechanisms work will locate this new technology in wider currents of thought. It is on much the same wider currents that the change in thinking about law that we address took place. This brings us to the common ancestor of the two revolutions.

1.4 TWO REVOLUTIONS WITH A COMMON ANCESTOR

Connections between two things in sequence do not necessarily mean that the later thing was caused by the one that came before, and a jurist who died in 1935 certainly was not the impetus behind recent advances in computer science. We aren't positing a connection between Holmes's jurisprudence and machine learning in that sense, nor is it our aim to offer an historical account of either law or computer science writ large. Our goal in this book is to explain how machine learning works by making an analogy to law—following Hempel's suggestion that connections across different domains can help people understand unfamiliar concepts.

Nonetheless, it is interesting to note an historical link between law and the mathematical sciences: the development of probabilistic thinking. According to philosopher of science Ian Hacking,

> [A]round 1660 a lot of people independently hit on the basic probability ideas. It took some time to draw these events together but they all happened concurrently. We can find a few unsuccessful anticipations in the sixteenth century, but only with hindsight can we recognize them at all. They are as nothing compared to the blossoming around 1660. The time, it appears, was ripe for probability.[21]

It's perhaps surprising to learn about the link between probability theory and law. In fact, the originators of mathematical probability were all either professional lawyers (Fermat, Huygens, de Witt) or the sons of lawyers (Cardano and Pascal).[22] At about the time Pascal formulated his famous wager about belief in God,[23] Leibniz thought of applying numerical probabilities to legal problems; he later called his probability theory "natural jurisprudence."[24] Leibniz was a law student at the time, though he is now better known for his co-invention of the differential calculus than for his law.[25] Leibniz developed his natural jurisprudence in order to reason mathematically about the weight of evidence in legal argument, thereby systematizing ideas that began with the Glossators of Roman Law in the twelfth century.[26] Law and probability theory both deal with evidence; the academic field of statistics is the science of reasoning about evidence using probability theory. Statistical theory for calculating the weight of evidence is now well understood.[27] Leibniz, if he were alive today, might find it interesting that judges are sometimes skeptical about statistics; but even where (as in a murder case considered by the Court of Appeal of England and Wales in 2010) courts have excluded statistical theory for some purposes, they have remained open to it for others.[28] Whether or not a given court in a given case admits statistical theory into its deliberations, the historical link between lawyers and probability remains.

As we said, though, our concern here is not with history as such. Probabilistic thinking is not only an historical link between law and the mathematical sciences. It is also the motive force behind the two modern revolutions that we are addressing. Machine learning (like any successful field) has many parents, but it's clear from any number of textbooks that probability theory is among the most important. As for Holmes, his particular interests in his formative years were statistics, logic, and the distinction between deductive and inductive methods of proof in science; he later wrote that "the man of the future is the man of statistics."[29] That Holmes's milieu was one of science and wide-ranging intellectual interests is a well-known fact of biography; his father was an eminent medical doctor and researcher, and the family belonged to a lively community of thinkers in Boston and Cambridge, the academic and scientific center of America at the time.

Less appreciated until recently is how wide and deep Holmes's engagement with that community and its ideas had been. Frederic R. Kellogg, in a magisterial study published in 2018 entitled *Oliver Wendell Holmes Jr. and Legal Logic*, has brought to light in intricate detail the groundings Holmes acquired in science and logic before his rise to fame as a lawyer. Holmes's interlocutors included the likes of his friends Ralph Waldo Emerson, Chauncey Wright, and the James brothers, William and Henry.[30] Holmes's attendance of the Lowell Lectures on logic and scientific induction delivered by Charles Peirce in 1866 exercised a particular and lasting influence on Holmes's thought.[31] Holmes spent a great deal of time as well with the writings of John Stuart Mill, including Mill's *A System of Logic, Ratiocinative and Inductive*. (He met Mill in London in 1866; they dined together with engineer and inventor of the electric clock Alexander Bain.[32]) Diaries and letters from the time record Holmes absorbed in conversation with these and other thinkers and innovators. Holmes eventually conceded his "Debauch on Philosophy" would have to subside if he ever were to become a practicing lawyer.[33]

Holmes clearly was interested in statistics, and statistics can be used to evaluate evidence in court. But Holmes's famous saying, to which we will return below, that law is nothing more than "prophecies of what the courts will do," points to a different use of probability theory: it points to prediction. Traditional statistical thinking is mostly concerned with making inferences about the truth of scientific laws and models, at least in so far as scientific models can be said to be "true." For example, an expert might propose an equation for the probability that a prisoner will reoffend, or that a defendant is guilty of a murder, and the statistician can estimate the terms in the equation and quantify their confidence. A different type of thinking was described by Leo Breiman in a rallying call for the nascent discipline of machine learning: he argued that prediction about individual cases is a more useful goal than inference about general rules, and that models should be evaluated purely on the accuracy of their predictions rather than on other scientific considerations such as parsimony or interpretability or consonance with theory.[34] For example, a machine learning programmer might build a device that predicts whether or not a prisoner will reoffend. Such a device can be evaluated on the accuracy of its predictions. True, society at large might insist that scrutiny be placed on the device to see whether its predictions come from sound considerations, whether using it comports with society's values, etc. But, in Breiman's terms, the programmer who built it should leave all that aside: the predictive accuracy of the device, in those terms, is the sole measure of its success. We will discuss the central role of prediction both in Holmes's thought and in modern machine learning in Chapters 5 and 6.

Time and again, revolutions in thought and method have coincided. Thomas Kuhn, among other examples in his *The Structure of Scientific Revolutions*, noted that a shift in thinking about what electricity is led scientists to change their experimental approach to exploring that natural phenomenon.[35] Later, and in a rather different setting, Peter Bernstein noted that changes in thinking about risk were involved in the emergence of the modern insurance industry.[36] David Landes considered the means by which societies measured time, how its measurement affected how societies thought about time, and how they thought about time in turn affected their behaviors and institutions.[37] The relations that interested these and other thinkers have been in diverse fields and have been of different kinds and degrees of proximity. A shift in scientific theory well may have direct impact on the program of scientific investigation; the transmission of an idea from theory to the marketplace might be less direct; the cultural and civilizational effects of new conceptions of the universe (e.g., conceptions of time) still less.[38]

Again, it is not our aim in this book to offer an historical account, nor a *tour d'horizon* of issues in philosophy of science or philosophy of law. Nor is it our aim to identify virtues or faults in the transformations we address. In computer science, it would be beside the point to "take sides" as between traditional algorithmic approaches to programming and machine learning. The change in technology is a matter of fact, not to be praised or criticized before its lineaments are accurately perceived. Nor, in law, is it to the present point to say whether it is good or bad that many jurists, especially since Holmes's time, have not kept faith with the formalist way of thinking about law. Battles continue to be fought over that revolution. We don't join those battles here.

What we do, instead, is propose that Holmes, in particular in his understanding of law as prediction formed from the search for patterns in experience, furnishes remarkably powerful analogies for machine learning. Our goal with the analogies is to explain the essence of how machine learning works. We believe that thinking about law in this way can help people understand machine learning as it is now—and help them think about where machine learning might go from here. People need both to grasp the state of the art and to think about its future, because machine learning gives rise to legal and ethical challenges that are difficult to recognize, even more to address, unless they do. Reading Holmes with machine learning in mind, we discern lessons about the challenges. Machine learning is a revolution in thinking, and it deserves to be understood much more widely and placed in a wider setting.

NOTES

1. As to the difference between "artificial intelligence" and "machine learning," see Prologue, pp. ix–x.
2. Kellogg (2018) 35, 72–87.
3. According to Brian Randell, a computer scientist writing in the 1970s, electronic digital computers had their origin in the late 1940s, but "[i]n most cases their developers were unaware that nearly all the important functional characteristics of these computers had been invented over a hundred years earlier by Charles Babbage," an English mathematician who had been "interested in the possibility of mechanising the computation and printing of mathematical tables." Randell, *The History of Digital Computers*, 12(11–12) IMA BULL. 335 (1976). Babbage (1791–1871) designed his "difference engine" around 1821, further to which see Swade, DIFFERENCE ENGINE: CHARLES BABBAGE AND THE

QUEST TO BUILD THE FIRST COMPUTER (2001). See also the timeline supplied by the Computer History Museum (Mountain View, California, USA): https://www.computerhistory.org/timeline/computers/. For a superb overview that locates these mechanical developments in the history of ideas, see *Historicizing the Self-Evident: An Interview with Lorraine Daston* (Jan. 25, 2000): https://lareviewofbooks.org/article/historicizing-the-self-evident-an-interview-with-lorraine-daston/.

4. Take for example proposals that the law confer legal personality on "robots": European Parliament, *Civil Law Rules on Robotics* resolution (Feb. 16, 2017): P8_TA(2017)0051, for a critique of which see Bryson, Diamantis & Grant (2017) 25 ART. INTEL. LAW 273. Deficiencies in the response to machine learning at political level owe, in no small part, to deficiencies in understanding and perspective as regards what machine learning is.

5. Chapter 10, pp. 114–119.

6. The Law Library of Congress, Global Legal Research Directorate (U.S.), supplies a worldwide survey by jurisdiction of legislation addressing AI, as well as overviews of work on the topic in international and regional intergovernmental organizations: REGULATION OF ARTIFICIAL INTELLIGENCE IN SELECTED JURISDICTIONS (Jan. 2019).

7. https://elitedatascience.com/machine-learning-impact (accessed July 27, 2019).

8. House of Lords Select Committee on Artificial Intelligence, Report (Apr. 16, 2018) pp. 63–94.

9. Executive Order 13859 of February 11, 2019, *Maintaining American Leadership in Artificial Intelligence.*

10. Ministry of Industry and Information Technology, *Three-Year Action Plan for Promoting Development of a New Generation Artificial Intelligence Industry* (2018–2020) (published Dec. 14, 2017), available in English translation (Paul Triolo, Elsa Kania & Graham Webster translators) at https://perma.cc/68CA-G3HL.

11. Communication from the Commission: *Artificial Intelligence for Europe*, COM (2018) 237 final (Apr. 25, 2018).

12. Compare these two subparagraphs of Section 238 in the U.S. National Defense Authorization Act for Fiscal Year 2019:

> 1. Any artificial system that performs tasks under varying and unpredictable circumstances without significant human oversight, or that can learn from experience and improve performance when exposed to data sets.

> An artificial system developed in computer software, physical hardware, or other context that solves tasks requiring human-like perception, cognition, planning, learning, communication, or physical action.

6 P.L. 115–232, Section 2, Division A, Title II, §238. The second paragraph displays the shortcoming of other circular definitions, but the first identifies, more helpfully, the relevance in AI of data and of learning from experience. See also the UN Secretary-General's initial take on the topic:

> 2. There is no universally agreed definition of artificial intelligence. The term has been applied in contexts in which computer systems imitate thinking or behavior that people associate with human intelligence, such as learning, problem-solving and decision-making.

Report of the Secretary-General, Current developments in science and technology and their potential impact on international security and disarmament efforts (July 17, 2018), A/73/177 p. 3 (¶5).

13. Daubert v. Merrell Dow Pharmaceuticals, Inc., 509 U.S. 579, 593, 113 S.Ct. 2786, 2797 (1993) (Blackmun J), quoting Hempel, PHILOSOPHY OF NATURAL SCIENCE 49 (1966): "[T]he statements constituting a scientific explanation must be capable of empirical test." For criticism, see Note, *Admitting Doubt: A New Standard for Scientific Evidence*, 123 HARV. L. REV. 2012, 2027 (2010); Susan Haack, *Of Truth, in Science and in Law*, 73 BROOK. L. REV. 985, 989–90 (2008). Hempel is not so prominently invoked in other jurisdictions but nevertheless has made the occasional appearance. See, e.g., Fuss v. Repatriation Commission [2001] FCA 1529 (Federal Court of Australia) (Wilcox J.) ¶42.

14. Hempel at 94 (emphasis added). For an overview of Hempel's contributions, see chapters in Fetzer (ed.) (2001).

15. We don't undertake further to justify analogy as a mode of argument or of explanation. It is with Hempel's insight on diverse connections that we rest our case in defense of analogies between seemingly disparate subjects. For a spirited defense of analogies *within* the law, see Frederick Schauer & Barbara A. Spellman, *Analogy, Expertise, and Experience*, 84 U. CHI. L. REV. 249 (2017) and esp. id. at 264–65 (on experience, expertise, and analogy).

16. An earlier generation of developments in AI inspired an earlier generation of work about how lawyers might use it. See, e.g., Zeleznikow & Hunter, BUILDING INTELLIGENT LEGAL INFORMATION SYSTEMS: REPRESENTATION AND REASONING IN LAW (1994); papers dating from 1987 to 2012 summarized in Bench-Capon et al., *A History of AI and Law in 50 Papers: 25 Years of the International Conference on AI and Law*, 20(3) ART. INTEL. LAW 215–310 (2012).

17. For example Zeleznikow & Hunter, 55 *ff*, 177–97.
18. See Prologue, p. x.
19. See for example Neil Sahota, *Will A.I. Put Lawyers Out of Business?* FORBES (Feb. 9, 2019). For a sober analysis, see the assessment by the Law Society of England and Wales, *Six Ways the Legal Sector Is Using AI Right Now* (Dec. 13, 2018): https://www.lawsociety.org.uk/news/stories/six-ways-the-legal-sector-is-using-ai/; and the Society's horizon-scanning report, *Artificial Intelligence and the Legal Profession* (May 2018). See also Remus & Levy's detailed task-by-task examination of AI and law practice: *Can Robots Be Lawyers? Computers, Lawyers, and the Practice of Law* (2016). For a sophisticated study of the use of a machine learning system to search legal materials, see Livermore et al., *Law Search as Prediction* (Nov. 5, 2018), Virginia Public Law and Legal Theory Research Paper No. 2018-61: https://ssrn.com/abstract=3278398.
20. See especially, addressing computational modeling to specific law practice tasks, Ashley (2017). See also many of the articles in the journal ART. INTEL. LAW (Springer); see also papers delivered at the biennial International Conference on Artificial Intelligence and Law (most recently, June 17–21, 2019, Montréal, Québec, Canada): https://icail2019-cyberjustice.com/. Journalism on practical applications of AI in law practice is vast. See, e.g., Bernard Marr, *How AI and Machine Learning Are Transforming Law Firms and the Legal Sector*, FORBES (May 23, 2018).
21. Hacking, The EMERGENCE OF PROBABILITY (1975) 11. See also Daston (1988) xi, 3–33.
22. Franklin, *Pre-history of Probability*, in Hájak & Hitchcock (eds.), OXFORD HANDBOOK OF PROBABILITY AND PHILOSOPHY (2016) 33–49.
23. Blaise Pascal, PENSÉES, Part II, especially §233. Translated by W.F. Trotter; introduction by T.S. Eliot (1908) pp. 65–69. Pascal (1632–1662) formulated the wager in notes that were published after his death. Hájek, in the entry on *Pascal's Wager* in Zalta (ed.), STANFORD ENCYCLOPEDIA OF PHILOSOPHY (2018), describes it as displaying "the extraordinary confluence of several important strands of thought" including "probability theory and decision theory, used here for almost the first time in history…".
24. Hacking, 11, 86. See further Artosi & Sartor, *Leibniz as Jurist* (2018) 641–63.
25. Richard T.W. Arthur, *The Labyrinth of the Continuum* in Antognazza (ed.) 275, 279–82. There has been, however, some interest recently in Leibniz's ideas about law, as to which see Livermore, *Rule by Rules* (forthcoming 2019) 2–4.
26. See Franklin (2001) 15–17, 26–27. Lorraine Daston thinks legal problems were at least as important as problems involving games of chance in shaping early thought about probability: Daston (1988) xvi, 6–7, 14–15.

27. See Fienberg & Kadane, *The Presentation of Bayesian Statistical Analyses in Legal Proceedings*, The STATISTICIAN 32 (1983); Dawid, *Probability and Statistics in the Law*, Research Report 243, Department of Statistical Science, University College London (May 2004).

28. Regina v. T. (2010) EWCA Crim 2439. An expert had testified about the probability of a match between an unusual pattern in a footprint at a murder scene and a criminal defendant's shoe. The Court of Appeal said that "[t]here is not a sufficiently reliable basis for an expert to be able to express an opinion based on the use of a mathematical formula" Id. at [86]. The Court further said that "an expert footwear mark examiner can... in appropriate cases use his experience to express a more definite evaluative opinion where the conclusion is that the mark 'could have been made' by the footwear. However no likelihood ratios or other mathematical formulae should be used in reaching that judgment" Id. at [95]. Redmayne et al., note that the Court of Appeal carefully limited its holding to footwear, acknowledging that probabilistic calculations in relation to DNA profiling and "possibly other areas in which there is a firm statistical base" might be admissible in criminal trials: Redmayne, Roberts, Aitken & Jackson, *Forensic Science Evidence in Question*, 5 CRIM. L. REV. 347, 351 (2011).

29. 10 HARV. L. REV. at 469.

30. Kellogg (2018), 48–49 and *passim*.

31. Id.

32. Id. 29, 46.

33. Id. 32.

34. Leo Breiman, *Statistical Modeling: The Two Cultures*, 16(3) STAT. SCI. 199–215 (2001). See also David Donoho, *50 Years of Data Science* (Sept. 18, 2015), from a presentation at the Tukey Centennial Workshop, Princeton, NJ.

35. Kuhn, *The Structure of Scientific Revolutions: II. The Route to Normal Science*, in (1962) 2(2) INTERNATIONAL ENCYCLOPEDIA OF UNIFIED SCIENCE at 17–18.

36. Bernstein was concerned especially with changes in thinking about risk in games of chance: Bernstein, AGAINST THE GODS: The REMARKABLE STORY OF RISK (1996) 11–22.

37. Landes, REVOLUTION IN TIME: CLOCKS AND THE MAKING OF THE MODERN WORLD (1983).

38. Which is not to say that relatively abstract changes in conception have less impact; they well may have more. As Landes described it, the mechanical clock was "one of the great inventions in the history of mankind—not in a class with fire and the wheel, but comparable to movable type in its revolutionary implications for cultural values, technological change, social and political organization, and personality... It is the mechanical clock that

made possible, for better or worse, a civilization attentive to the passage of time, hence to productivity and performance" Id. at 6–7.

CHAPTER 2

Getting Past Logic

In law, as we saw in Chapter 1, the contrast between formalism and inductivism was evident practically from the moment jurists began to consider that logic might not explain everything about the law. The contrast continues to define lines that run through legal studies and judicial politics, especially in the United States and also to an extent in other common law jurisdictions. The lines are readily discernible. Volumes of literature and on-going disputes are gathered on one side or the other. People who think about and practice law have identified themselves in adversarial terms by reference to which side of those lines they stand on. In computing, as we also noted in Chapter 1, the lines are not drawn in quite such clear relief. They are certainly not the reference point for the intellectual identity of opposing camps of computer scientists. The conceptual shift that underpins the emergence of machine learning has had enormous impact, but it has not been an object of sustained discourse, much less of pitched ideological battle. Computer scientists thus, perhaps, enjoy a certain felicity in their professional relations, but they also are probably less alert to the distinction that machine learning introduces between their present endeavors and what they were doing before.

We will examine in the three chapters immediately after this the ingredients that go into making machine learning so different from traditional algorithm-based programming. The ingredients are data (Chapter 3), pattern finding (Chapter 4), and prediction (Chapter 5). Before examining these, we wish to consider further the contrast between machine learning and what came before. The rise of legal realism was explicitly a challenge

© The Author(s) 2020
T. D. Grant and D. J. Wischik, *On the path to AI*,
https://doi.org/10.1007/978-3-030-43582-0_2

to what came before in law, and, so, the contrast was patent. With the emergence of induction-driven machine learning, the contrast ought to be no less clear, but people continue to miss it. Getting past logic is necessary, if one is to get at what's new about machine learning—and at why this kind of computing presents special challenges when it comes to values in society at large.

2.1   FORMALISM IN LAW
AND ALGORITHMS IN COMPUTING

Legal formalists start with the observation, to which few would object, that law involves rules. They identify the task in law, whether performed by a lawyer or a judge, to be that of applying the rules to facts, again in itself an unobjectionable, or at least unremarkable, proposition. Where formalism is distinctive is in its claim that these considerations supply a complete understanding of the law. "Legal reasoning," said the late twentieth century critical legal scholar Roberto Unger, "is formalistic when the mere invocation of rules and deduction of conclusions from them is believed sufficient for every authoritative legal choice."[1] An important correlate follows from such a conception of the law.[2] The formalists say that, if the task of legal reasoning is performed correctly, meaning in accordance with the logic of the applicable rules, the lawyer or judge reaches the correct result. The result might consist in a judgment (adopted by a judge and binding on parties in a dispute) or in a briefing (to her client by a lawyer), but whatever the forum or purpose, the result comes from a logical operation, not differing too much from the application of a mathematical axiom. In the formalists' understanding, it thus follows that the answer given to a legal question, whether by a lawyer or by a judge, is susceptible of a logical process of review. An erroneous result can be identified by tracing back the steps that the lawyer or judge was to have followed and finding a step in the operation where that technician made a mistake. Why a correct judgment is correct thus can be explained by reference to the rules and reasoning on which it is based; and an incorrect one can be diagnosed much the same way.

In Oliver Wendell Holmes, Jr.'s day, though legal formalism already had a long line of distinguished antecedents such as Blackstone (whom we quoted in our Prologue), one contemporary of Holmes, C. C. Langdell, Dean and Librarian of the Harvard Law School, had come to be specially associated with it. Holmes himself identified the Dean as arch-exponent of this mode of legal reasoning. In a book review in 1880, he referred to Langdell, who was a friend and colleague, as "the greatest living legal theologian."[3] The compliment was a back-handed one when spoken among self-respecting rationalists in the late nineteenth century. In private correspondence around the same time, Holmes called Langdell a jurist who "is all for logic and hates any reference to anything outside of it."[4] A later scholar, from the vantage of the twenty-first century, has suggested that Langdell was less a formalist than Holmes and others made him out to be but nevertheless acknowledges the widespread association and the received understanding: "[l]egal formalism [as associated with Langdell] consisted in the view that deductive inference from objective, immutable legal principles determines correct decisions in legal disputes."[5]

Whether or not Langdell saw that to be the only way the law functions, Holmes certainly did not, and the asserted contrast between the two defined lines which remain familiar in jurisprudence to this day. In his own words, Holmes rejected "the notion that the only force at work in the development of the law is logic."[6] By this, he did *not* mean that "the principles governing other phenomena [do not] also govern the law."[7] Holmes accepted that logic plays a role in law: "[t]he processes of analogy, discrimination, and deduction are those in which [lawyers] are most at home. The language of judicial decision is mainly the language of logic."[8] That deductive logic and inductive reasoning co-exist in law may already have been accepted, at least to a degree, in Holmes's time.[9] What Holmes rejected, instead, was "the notion that a [legal system]... can be worked out like mathematics from some general axioms of conduct."[10] It was thus that Holmes made sport of a "very eminent judge" who said "he never let a decision go until he was absolutely sure that it was right" and of those who treat a dissenting judgment "as if it meant simply that one side or the other were not doing their sums right, and if they would take more trouble, agreement inevitably would come."[11] If the strict formalist thought that all correctness and error in law are readily distinguished and their points of origin readily identified, then Holmes thought that formalism as legal theory was lacking.

The common conception of computer science is analogous to the formalist theory of law. Important features of that conception are writing a problem description as a formal specification; devising an algorithm, i.e. a step-by-step sequence of instructions that can be programmed on a computer; and analyzing the algorithm, for example to establish that it correctly solves the specified problem. "The term *algorithm* is used in computer science to describe a … problem-solving method suitable for implementation as a computer program. Algorithms are the stuff of computer science: they are central objects of study in the field."[12] In some areas the interest is in devising an algorithm to meet the specification. For example, given the problem statement *Take a list of names and sort them alphabetically*, the computer scientist might decompose it recursively into *to sort a list, first sort the first half, then sort the second half, then merge the two halves*, and then break these instructions down further into elementary operations such as *swap two particular items in the list*. In other areas the interest is in the output of the algorithm. For example, given the problem statement *Forecast the likely path of the hurricane*, the computer scientist might split a map into cells and within each cell solve simple equations from atmospheric science to predict how wind speed changes from minute to minute. In either situation, the job of the computer scientist is to codify a task into simple steps, each step able to be (i) executed on a computer, and (ii) reasoned about, for example to debug why a computer program has generated an incorrect output (i.e. an incorrect result). The steps are composed in source code, and scrutinizing the source code can disclose how the program worked or failed. Success and failure are ready to see. The mistakes that cause failure, though sometimes frustratingly tangled in the code, are eventually findable by a programmer keen enough to find them.

2.2   GETTING PAST ALGORITHMS

Machine learning however neither works like algorithmic code nor is to be understood as if it were algorithmic code. Outputs from machine learning are not effectively explained by considering only the source code involved. Kroll et al., whom we will consider more closely in a moment, in a discussion of how to make algorithms more accountable explain:

> Machine learning… is particularly ill-suited to source code analysis because it involves situations where the decisional rule itself emerges automatically from the specific data under analysis, sometimes in ways that no human can explain. In this case, source code alone teaches a reviewer very little, since the code only exposes the machine learning method used and not the data-driven decision rule.[13]

In machine learning, the job of the computer scientist is to assemble a training dataset and to program a system that is capable of learning from that data. The outcome of training is a collection of millions of fine-tuned parameter values that configure an algorithm. The algorithms that computer scientists program in modern machine learning are embarrassingly simple by the standards of classic computer science, but they are enormously rich and expressive by virtue of their having millions of parameters.

The backbone of machine learning is a simple method, called *gradient descent*.[14] It is through gradient descent that the system arrives at the optimum settings for these millions of parameters. It is how the system achieves its fine-tuning. To be clear, it is not the human programmer who fine tunes the system; it is a mathematical process that the human programmer sets in motion that does the fine-tuning. Thus built on its backbone of gradient descent, machine learning has excelled at tasks such as image classification and translation, tasks where formal specification and mathematical logic have not worked. These achievements justify the encomia that this simple method has received. "Gradient descent can write code better than you."[15] After training, i.e. after configuring the algorithm by setting its parameter values, the final stage is to invoke the algorithm to make decisions on instances of new data. It is an algorithm that is being invoked, in the trivial sense that it consists of simple steps which can be executed on a computer; but its behavior cannot be understood by reasoning logically about its source code, since its source code does not include the learnt parameter values.

Moreover, it is futile to try to reason logically about the algorithm even given all the parameter values. Such an analysis would be as futile as analyzing a judge's decision from electroencephalogram readings of her brain. There are just too many values for an analyst to make sense of. Instead, machine-learnt algorithms are evaluated empirically, by measuring how they perform on test data. Computer scientists speak of "black-box" and "white-box" analysis of an algorithm. In white-box analysis we consider the internal structure of an algorithm, whereas in black-box analysis we consider only its inputs and outputs. Machine-learnt algorithms are evaluated purely on the basis of which outputs they generate for which inputs, i.e. by black-box analysis. Where computer scientists have sought to address concerns about discrimination and fairness, they have done so with black-box analysis as their basis.[16] In summary, a machine learning "algorithm" is better thought of as an opaque embodiment of its training dataset and evaluation criterion, not as a logical rules-based procedure. Problems with which machine learning might be involved (such as unfair discrimination) thus are not to be addressed as if it were a logical rules-based procedure.

2.3   THE PERSISTENCE OF ALGORITHMIC LOGIC

Yet people continue to address machine learning as if it were just that—a logical rules-based procedure not different in kind from traditional computer programming based on algorithmic logic. This inadequate way of addressing machine learning—addressing it as though the source code of an algorithm is responsible for producing the outputs—is not limited to legal discourse. It is however very much visible there. Formal, algorithmic descriptions of machine learning are ubiquitous in legal literature.[17] The persistence of algorithmic logic in descriptions of how computers work is visible even among legal writers who otherwise acknowledge that machine learning is different.[18]

Even Kroll et al., who recognize that machine learning "is particularly ill-suited to source code analysis," still refer to "a machine [that] has been 'trained' through exposure to a large quantity of data and *infers a rule from the patterns it observes.*"[19] To associate machine learning with "a rule from the patterns it observes" will lead an unwary reader to conclude that the machine has learnt a cleanly stated rule in the sense of law or of decision trees. In fact, the machine has done no such thing. What it has done is find a pattern which is "well beyond traditional interpretation," these being the much more apt words that Kroll et al. themselves use to acknowledge the opacity of a machine learning mechanism.[20]

Kroll and his collaborators have addressed at length the challenges in analyzing the computer systems on which society increasingly relies. We will turn in Chapters 6–8 to extend our analogy between two revolutions to some of the challenges. A word is in order here about the work of Kroll et al., because that work highlights both the urgency of the challenges and the traps that the persistence of algorithmic logic presents.

There are many white-box tools for analyzing algorithms, for example based on mathematical analysis of the source code. Kroll et al. devote the bulk of their paper on *Accountable Algorithms* (2017) to white-box software engineering tools and to the related regulatory tools that can be used to ensure accountability. The persistence of algorithmic logic here, again, may lead to a trap: the unwary reader thinks machine learning algorithms are per se algorithms; here are tools for making algorithms accountable; *therefore we can make machine learning accountable*. Kroll et al. mark the trap with a warning flag, but it is a rather small one. They concede in a footnote that all the white-box techniques they discuss simply do not apply to machine learning mechanisms and that the best that can be done is to regulate the *decision* to use machine learning: "Although some machine learning systems produce results that are difficult to predict in advance and well beyond traditional interpretation, the choice to field such a system instead of one which can be interpreted and governed is itself a decision about the system's design."[21] This isn't saying how to understand machine learning better. It's saying not to use machine learning. The result, if one were to follow Kroll et al., would be to narrow the problem set to a much easier question—what to do about systems that use only traditional algorithmic logic.

Indeed, it is the easier question that occupies most of Kroll et al.'s *Accountable Algorithms*. Forty-five pages of it discuss white-box analysis that doesn't apply to machine learning systems. Eighteen pages then consider black-box analysis. An exploration of black-box analysis is more to the point—the point being to analyze machine learning.

But a trap is presented there as well. The pages on black-box analysis are titled "Designing algorithms to ensure fidelity to substantive policy choices." Black-box analysis by definition is agnostic as to the design of the "algorithms" that are producing the results it analyzes. Black-box analysis is concerned with the output of the system, not with the inner workings that generate the output. To suggest that "[d]esigning algorithms" the right way will "ensure fidelity" to some external value is to fall into the algorithmic trap. It is to assume that algorithmic logic is at work, when the real challenge involves machine learning systems, the distinctive feature of which is that they operate outside that logic. Black-box analysis, which Kroll et al. suggest relies upon the design of an algorithm, in fact works just as well in analyzing decisions made by a flock of dyspeptic parrots.

Kroll in a later paper expands the small flag footnote and draws a distinction between systems and algorithms.[22] As Kroll uses the words, the *system* includes the human sociotechnical context—the power dynamics and the human values behind the design goals, and so on—whereas the *algorithm* is the technical decision-making mechanism embedded in the system. It is the "system," in the sense that he stipulates, that mainly concerns him.[23] Kroll argues that a machine learning "system" is necessarily scrutable, since it is a system built by human engineers, and human choices can always be scrutinized. But, once more, this is an observation that would apply just as much if the engineers' choice was to use the parrot flock. It is the essence of the black-box that we know only what output it gives, whether a name, a color, a spatial coordinate, or a squawk. We don't know how it arrived at the output. In so far as Kroll addresses machine learning itself, he does not offer tools to impart scrutability to it but, instead, only this: "the question of how to build effective white-box testing regimes for machine learning systems is far from settled."[24] To say that white-box testing doesn't produce "settled" answers to black-box questions is an understatement. And the problem it understates is the very problem that activists and political institutions are calling on computer scientists to address: how to test a machine learning system to assure that it does not have undesirable effects. What one is left with, in talking about machine learning this way, is a hope: namely, a hope that machine learning, even though it may be the most potent mechanism yet devised for computational operations, will not be built into "systems" by the many individuals and institutions who stand to profit from its use. Exploration of white-box, logical models of scrutability reveals little or nothing about machine learning. Insistence on such exploration only highlights the persistence of algorithmic logic notwithstanding the revolution that this technology represents.

Many accounts of machine learning aimed at non-specialists display these shortcomings. Lawyers, as a particular group of non-specialist, are perhaps particularly susceptible to misguided appeals to logical models of computing. It is true that statutory law has been compared to computer source code[25]; lawyers who are at heart formalists may find the comparison comforting.[26] It is also true that an earlier generation of software could solve certain fairly hard legal problems, especially where a statutory and regulatory regime, like the tax code, is concerned. Machine learning, however, is not limited in its applications to tasks that, like calculating a tax liability, are themselves algorithmic (in the sense that a human operator can readily describe the tasks as a logical progression applying fixed rules to given facts). Computer source code is not the characteristic of machine learning that sets it apart from the kind of computing that came before. Lawyers must let go the idea that logic—the stepwise deduction of solutions by applying a rule—is what's at work in the machine learning age. Herein, we posit, reading Holmes has a salutary effect.

Holmes made clear his position by contrasting it against that taken by jurists of what we might, if anachronistically, call the algorithmic school, that is to say the formalists. In *Lochner v. New York*, perhaps his most famous dissent, Holmes stated, "General propositions do not decide concrete cases."[27] This was to reject deductive reasoning in plain terms and in high profile. Whether or not we think that is a good way to think about law,[28] it is precisely how we must think if we are to understand machine learning; machine learning demands that we think beyond logic. Computer scientists themselves, as much as lawyers and other laypersons, ought to recognize this conceptual transition, for it is indispensable to the emergence of machine learning which is now transforming their field and so much beyond.

With the contrast in mind between formal ways of thinking about law and about computing, we now will elaborate on the elements of post-logic thinking that law and computing share: data, pattern finding, and prediction.

NOTES

1. Unger (1976) 194.
2. Unger's description is not the only or necessarily the definitive description of legal formalism. A range of nuances exists in defining legal formalism. Unger's pejorative use of the word "mere," in particular, may well be jettisoned: to say that logical application of rules to facts is the path to legal results by no means entails that the cognitive operations involved—and opportunities for disagreement—are trivial. Dijkstra's caution that only very good mathematicians are ready for the logical rigors of (traditional) computing merits an analogue for lawyers.
3. Holmes, *Book Notice Reviewing a Selection of Cases on the Law of Contracts, with a Summary of the Topics Covered by the Cases, By C.C. Langdell,* 14 AM. L. REV. 233, 234 (1880).
4. Letter from Holmes to Pollock (Apr. 10, 1881), reprinted De Wolfe Howe (ed.) (1942) 17. Though Holmes's skepticism over the formalist approach most famously targeted Langdell, Walbridge Abner Field, Chief Justice of the Massachusetts Supreme Judicial Court, also fell into the sights. See Budiansky (2019) 199–200.
5. Kimball (2009) 108 and references to writers id. at 108 n. 134. See also Cook 88 N.D. L. REV. 21 (2012); Wendel, 96 CORN. L. REV. 1035, 1060–65, 1073 (2011). Langdell's undoubted contribution—the case method of teaching law—provoked rebellion by his students for the very reason that it did *not* start with clear rules: see Gersen, 130 HARV. L. REV. 2320, 2323 (2017). Cf. Grey, *Langdell's Orthodoxy*, 45 U. PITT. L.

Rev. 1 (1983), who drew attention to the rigor and advantages of classic formalism.

Richard Posner, to whose understanding of Holmes's legal method we will return later below (Chapter 10, p. 119), describes logic as central to legal reasoning of the time, though "common sense" was also involved:

> The task of the legal scholar was seen as being to extract a doctrine from a line of cases or from statutory text and history, restate it, perhaps criticize it or seek to extend it, all the while striving for 'sensible' results in light of legal principles and common sense. Logic, analogy, judicial decisions, a handful of principles such as *stare decisis*, and common sense were the tools of analysis.

Posner, 115(5) Harv. L. Rev. 1314, 1316 (2002).

6. 10 Harv. L. Rev. at 465.
7. Id. For a consideration of the role of Langdell's logic in Holmes's conception of law, see Brown & Kimball, 45 Am. J. Leg. Hist. 278–321 (2001).
8. 10 Harv. L. Rev. at 465–66.
9. Which is not to say that Holmes necessarily would have agreed with other jurists' understanding of *how* they co-exist. Edward Levi, in his text on legal reasoning, said this: "It is customary to think of case-law reasoning as inductive and the application of statutes as deductive," Levi, An Introduction to Legal Reasoning (1949) 27. Common law courts have described their function in applying rules derived from past cases as involving both kinds of reasoning—induction to derive the rules; deduction from a rule thus derived. See, e.g., Skelton v. Collins (High Court of Australia, Windeyer J., 1966) 115 CLR 94, 134; Home Office v. Dorset Yacht Co Ltd. [1970] A.C. 1004, 1058–59 (House of Lords, Diplock LJ). Cf. Norsk Pacific Steamship et al. v. Canadian National Railway Co., 1992 A.M.C. 1910, 1923 (Supreme Court of Canada, 1992, McLachlin J.); In the Matter of Hearst Corporation et al. v. Clyne, 50 N.Y.2d 707, 717, 409 N.E.2d 876, 880 (Court of Appeals of New York, 1980, Wachtler J.). Cf. Benjamin Cardozo, The Nature of the Judicial Process (1921) 22–23. Levi's contrast—between induction for the common law and deduction for rules such as statutes contain—has been noted in connection with computer programming: see, e.g., Tyree (1989) 131.
10. 10 Harv. L. Rev. at 465.
11. Id.
12. Robert Sedgewick, Algorithms (4th ed.) (2011) Section 1.
13. Kroll et al., *Accountable Algorithms*, U. Pa. L. Rev. at 638 (2017).
14. For a formal description of gradient descent, and some of its variants, see e.g. Bishop, Pattern Recognition and Machine Learning

(2007) Section 5.2.4. Gradient descent is attributed to the French mathematician Augustin-Louis Cauchy (1789–1857) in a work published in 1847 (see Claude Lemaréchal, *Cauchy and the Gradient Method* (2012), DOCUMENTAL MATH. p. 251). Gradient descent when applied to finding parameter values for a neural network is known as *back propagation*.

15. Tweet by @karpathy, 1.56 p.m., Aug. 4, 2017. https://twitter.com/karpathy/status/893576281375219712? See further, e.g., Murphy (2012) 247, 445.

16. Kroll et al., *supra* n. 76, at 650–51, 685–87. The tests for discrimination in machine learning algorithms described there, which come from Cynthia Dwork et al., *Fairness Through Awareness*, ITCS CONF. PROC. (3rd) (2012), derive from a black-box definition.

17. See for example works cited by Barocas & Selbst, *Big Data's Disparate Impact*, 104 CAL. L. REV. 671, 674 n. 11. See also Alan S. Gutterman, *Glossary of Computer Terms*, in BUSINESS TRANSACTIONS SOLUTIONS § 217:146 (Jan. 2019 update) ("the machine merely follows a carefully constructed program"); Chessman, 105 CAL. L. REV. 179, 184 (2017) ("Evidence produced by computer programs arguably merits additional scrutiny… because the complexity of computer programs makes it difficult… to detect errors"); Gillespie in Gillespie et al. (eds.) (2014) 167, 192 ("algorithmic logic… depends on the proceduralized choices of a machine designed by human operators to automate some proxy of human judgment"), and the same as quoted by Carroll, *Making News: Balancing Newsworthiness and Privacy in the Age of Algorithms*, 106 GEO. L.J. 69, 95 (2017). As we will observe below, algorithmic logic persists in legislation and regulation as well: see Chapter 6, p. 70.

18. See for example Roth, 126 YALE L. J. 1972, 2016–2017 (2017) who says, "[A] machine's *programming*… could cause it to utter a falsehood by design" (emphasis added). Another example is found in Barocas & Selbst, 104 CAL. L. REV. at 674 (2016), who, like Roth, understand that "the data mining process itself" is at the heart of machine learning, but still say that "[a]lgorithms could exhibit these [invidious] tendencies even if they have not been manually programmed to do so…", a formulation which accounts for programs that are not written in the conventional way, but which still does not escape the gravity of the idea of source code (i.e., that which is "programmed"). Benvenisti has it on the mark, when he cites Council of Europe Rapporteur Wagner's observation that "provision of all of the source codes… may not even be sufficient." Benvenisti, *Upholding Democracy and the Challenges of New Technology: What Role for the Law of Global Governance?* 29 EJIL 9, 60 n. 287 (2018) quoting, inter alia, Wagner, Rapporteur, Committee of Experts on Internet Intermediaries (MSI-NET, Council of Europe), *Study on the Human Rights Dimensions of Algorithms* (2017) 22.

19. Kroll et al., *supra* n. 76, at 679 (emphasis added).
20. Id., n. 9. Others have used the same terms, similarly running the risk of conflating the output of machine learning systems with "rules." For example distinguished medical researchers Ziad Obermeyer & Ezekiel J. Emanuel, who have written cogent descriptions of how machine learning works, refer to machine learning as "approach[ing] problems as a doctor progressing through residency might: *by learning rules from data*" (emphasis added). Obermeyer & Ezekiel, *Predicting the Future—Big Data, Machine Learning, and Clinical Medicine,* 375(13) NEJM 1216 (Sept. 29, 2016).
21. Kroll et al. at n. 9. The proviso "*some* machine learning systems" acknowledges that at least some machine learning systems such as neural networks are well beyond traditional interpretation. True, some simple systems labeled "machine learning" produce easily understood rules and are therefore amenable to white-box analysis. One hears sales pitches that use the term "machine learning" to refer to basic calculations that could be performed in a simple Excel spreadsheet. These are not the systems making the emerging machine learning age, or its challenges, distinctive.
22. Kroll (2018) Phil. Trans. R. Soc. A 376.
23. Kroll (2018) indeed is about systems not algorithms—e.g., it uses "algorithm" and related words 16 times, and "system" and related words 235 times.
24. Id., p. 11. There have been attempts to find white-box approaches for understanding neural networks, notably Shwartz-Ziv & Tishby (2017), but such theory is not widely accepted: Saxe et al. (2018).
25. See esp. Lessig's extended analogy: Lessig, Code and Other Laws of Cyberspace (1999) 3–8, 53. See also Tyree (1989) 131.
26. For an analogy from psychology positing that legal formalism comes before supposedly "later stages of cognitive and moral development," see Huhn, 48 Vill. L. Rev. 305, 318–39 (2003). We venture no opinion as to the whether Huhn's analogy is convincing, either for law or for computer science.
27. *Lochner v. New York* , 198 U.S. 45, 76 (1905) (Holmes, J., dissenting).
28. Posner observes jurists have disputed formalism for two thousand years: Posner (1990) 24–25. The contrast between formalism and other types of legal analysis, whenever the latter arose and whatever name they go by ("positivism," "realism," etc.), is at the heart of much of the intellectual—and ideological—tension in American lawyering, judging, and law scholarship. It is visible in international law as well. See Prologue, pp. xii–xiii, n. 17. As we stated in Chapter 1, we address the contrast for purposes of analogy to machine learning, not to join in a debate over the merits of formalism or its alternatives in law.

CHAPTER 3

Experience and Data as Input

We are entering the era of big data. For example, there are about 1 trillion web pages; one hour of video is uploaded to YouTube every second, amounting to 10 years of content every day; the genomes of 1000s of people, each of which has a length of 3.8 × 10⁹ base pairs, have been sequenced by various labs; Walmart handles more than 1M transactions per hour and has databases containing more than 2.5 petabytes (2.5 × 10¹⁵) of information; and so on.

> *Kevin P. Murphy,* MACHINE LEARNING: A PROBABILISTIC PERSPECTIVE, *p. 1,* © *2012 Massachusetts Institute of Technology, published by The MIT Press*

The life of the law has not been logic; it has been experience.

> *Oliver Wendell Holmes, Jr.,* THE COMMON LAW *(1881) p. 1*

Holmes, when he articulated a way of thinking about law that departed from the prevalent deductive formalism of his day, traced an outline recognizable in twenty-first century computer science. The nineteenth century understanding of legal reasoning, which Holmes thought at best incomplete, had been that the law, like an algorithm, solves the problems given to it in a stepwise, automatic fashion. A well-written law applied by a technically competent judge leads to the correct judgment; a bad judgment owes to a defect in the law code or in the functioning of the judge. Holmes had a contrasting view. In Holmes's view, the judge considers a body of information, in the form of existing decisions and also, though the judge might not admit it, in the form of human experience at

© The Author(s) 2020
T. D. Grant and D. J. Wischik, *On the path to AI*,
https://doi.org/10.1007/978-3-030-43582-0_3

large, and in that body discerns a pattern. The pattern is the law itself. As computer science has developed from algorithm to machine learning, it, too, has departed from models that find satisfactory explanation in formal proof. In machine learning, the input is data, as in the law in Holmes's view the input is experience; and, in both, the task to be performed upon a given set of inputs is to find patterns therein. Thus in two different fields at different times, a transition has occurred from logic applied under fixed rules to a search for patterns.

In the present chapter, we consider more closely the *inputs*—experience and data; in Chapter 4 we will consider how, in both law and machine learning, patterns are found to make sense of the inputs; and in Chapter 5 we turn to the *outputs*, which, as we will see, are *predictions* that emerge through the search for pattern.

3.1   Experience Is Input for Law

To what materials does one turn, when one needs to determine the rules in a given legal system? Holmes had a distinctive understanding of how that question is in fact answered. In *The Common Law*, which was published sixteen years before *The Path of the Law*, Holmes started with a proposition that would join several of his aphorisms in the catalogue of jurists' favorites: "The life of the law has not been logic; it has been experience."[1] This proposition was further affirmation of Holmes's view that logic, on its own, only gets the jurist so far. More is needed if a comprehensive understanding of the legal system is to be reached. Holmes proceeded:

> The felt necessities of the time, the prevalent moral and political theories, intuitions of public policy, avowed or unconscious, and even the prejudices which judges share with their fellow-men, have had a good deal more to do than syllogism in determining the rules by which men should be governed. The law embodies the story of a nation's development through many centuries, and it cannot be dealt with as if it contained only the axioms and corollaries of a book of mathematics.[2]

We see here again the idea, recurrent in Holmes's writing, that law is not about formal logic, that it is not like mathematics. We also see an expansion upon that idea, for here Holmes articulated a theory of where law does come from. Where Holmes rejected syllogism—dealing with the law through "axioms and corollaries"—he embraced in its place the systematic understanding of experience. The experience most relevant to the law consists of the recorded decisions of organs having authority over the individual or entity subject to a particular legal claim—judgments of courts, laws adopted by parliaments, regulations promulgated by administrative bodies, and so on.

Holmes understood experience as wider still, however, for he did not invoke only formal legal texts but also "prevalent moral and political theories, intuitions of public policy... even the prejudices which judges share with their fellow-men."[3] The texts of law, for Holmes, were part of the relevant data but taken on their own not enough to go on.

In response to Holmes's invocation of sources such as political theory and public policy, one might interject that, surely *some* texts have undoubted authority, even primacy, over a given legal system—for example, a written constitution, to give the surest case. In Holmes's view, however, one does not reach the meaning even of a constitution through logic alone. It is to history there as well that Holmes would have the lawyer turn:

> The provisions of the Constitution are not mathematical formulas that have their essence in form, they are organic, living institutions transplanted from English soil. Their significance is vital, not formal; it is to be gathered not simply by taking the words and a dictionary but by considering their origin and the line of their growth.[4]

That Holmes was a keen legal historian is not surprising.[5] When he drew attention to "a nation's development through many centuries," this was directly to his purpose and to his understanding of the law. For Holmes, experience in its broadest sense goes into ascertaining the law.

3.2   DATA IS INPUT FOR MACHINE LEARNING

As we suggested in Chapter 2, a common misperception is that machine learning describes a type of decision-making algorithm: that you give the machine a new instance to decide, that it does some mysterious algorithmic processing, and then it emits an answer. In fact, the clever part of machine learning is in the *training phase*, in which the machine is given a dataset, and a learning algorithm converts this dataset into a digest. Holmes talked about a jurist processing a rich body of experience from which a general understanding of the law took form. In the case of modern machine learning, the "experience" is the data; the general understanding is in the digest, which is stored as millions of finely-tuned parameter values. We call these values "learnt parameters." The learnt parameters are an analogue (though only a rather rough one) to the connection map of which neurons activate which other neurons in a brain.

The training dataset—the "experience" from which the system learns—is of vital importance in determining the shape that the system eventually assumes. Some further words of detail about the training dataset thus are in order.

Computer scientists describe the training dataset in terms of *feature* variables and *outcome* variables. To see how these terms are used, let us take an example of how we might train a machine to classify emails as either spam or not spam. The outcome variable in our example is the label "spam" or "not-spam." The feature variables are the words in the email. The training dataset is a large collection of emails—together with human-annotated labels (human-annotated, because a twenty-first century human, unlike an untrained machine, knows spam when he sees it). In the case of legal experience, the facts of a case would be described as feature variables, and the judgment would be described as an outcome variable.

There is a subfield of machine learning, so-called "unsupervised" machine learning, in which the dataset consists purely of feature variables without any outcome variables. In other words, the training dataset does not include human-annotated labels. The learning process in that kind of machine learning consists in finding patterns in the training dataset. That kind of machine learning—unsupervised machine learning—corresponds to Holmes's broader conception of experience as including "prevalent moral and political theories" and the whole range of factors that might shape a jurist's learning. Classifications are not assigned to the data a priori through the decision of some formal authority. They are instead discerned in the data as it is examined.

After the machine has been trained, i.e. after the machine has carried out its computations and thus arrived at learnt parameter values from the training dataset, it can be used to give answers about new cases. We present the machine at that point with *new* feature variables (the words in a new email, which is to say an email not found in the training dataset), and the machine runs an algorithm that processes these new feature variables together with the learnt parameters. By doing this, the machine produces a predicted outcome—in our example, an answer to the question whether that new email is to be labeled "spam" or "not-spam." We will consider further below (in Chapter 5)[6] the predictive character of machine learning, which is shared by Holmes's idea of law.

Data, especially "big data," is the grist for machine learning. The word *data* is apt. It comes from the Latin datum, "that which is given," the past participle of *dare*, "to give." The dataset used to train a machine learning system (whether or not classifications are assigned to the data in the dataset a priori) is treated as a *given* in this sense: the dataset is stipulated to be the "ground truth"—the source of authority, however arbitrary. A machine learning system doesn't question or reason about what it is learning. The predictions are nothing more than statements in the following form: "such and such a new case is likely to behave similarly to other similar cases that belong to the dataset that was used to train this machine." It was an oft-noted inclination of Holmes's to take as a given the experience from which law's patterns emerge.[7] The central objection commonly voiced about Holmes's legal thinking—that he didn't care about social or moral values—would apply by analogy to the predictions derived from data. We will explore this point and its implications in Chapters 6–10 below.

In typical machine learning, the training dataset thus is assembled beforehand, the parameters are learnt, and then the trained machine is put to use. Holmes's concept of law follows a similar path. The collected experience of society (including its written legal texts) may be likened to the training dataset. The learnt experience of a jurist may be likened to the parameter values in a machine learning system. The jurist is presented new questions, just as the machine (after training has produced the learnt parameters) is presented new feature variables, and, from both, outputs are expected.

Jurists will naturally keep accumulating experience over time, both from the cases they have participated in and from other sources. In a particular variant of machine learning, a machine likewise can undergo incremental training once it has been deployed. This is described as *online learning*, denoting the idea that the machine has "gone online" (i.e., become operational) and continues to train. On grounds of engineering simplicity it's more common, so far, to train the machine and then deploy it without any capability for online learning.[8]

There is perhaps an aspect of Holmes's understanding of the law that does not (yet) have any counterpart in machine learning, even its online variant: a legal decision is made *in anticipation* of how it will be used as input for future decisions. An anticipatory aspect is not present in machine learning in its present state of the art. We will explore this idea in Chapter 9.

3.3   THE BREADTH OF EXPERIENCE AND THE LIMITS OF DATA

Another distinction is that the experience Holmes had in mind is considerably broader than the typical training datasets used in machine learning, and it is less structured. The machine learning system is constrained to receive inputs in simple and rigid formats. For example, a machine receives an input in the form of an image of prespecified size or a label from a prespecified (relatively) small set of possibilities; its output is an image or a label of the same form. The tasks that machine learning can handle, in the present state of the art, are those where the machine is asked to make a prediction about things that are new to the machine, but whose newness does not exceed the parameters of the data on which the machine was trained. Machine learning is limited in this respect. It is limited to data in a particular sense—data as a structured set of inputs; whereas the experience in which jurists find the patterns of law is of much wider provenance and more varied shape.

Machine learning, however, is catching up. There is ongoing research on how to incorporate broad knowledge bases into machine learning systems, for example to incorporate knowledge about the world obtained from Wikipedia. Any very large and highly variegated dataset could be an eventual training source, if machine learning gets to that goal. The case reports of a national legal system would be an example, too, of the kind of knowledge base that could be used to train a machine learning system. To the extent that computer science finds ways to broaden the data that can be used to train a machine learning system, the data training set will come that much more to resemble Holmes's concept of experience as the basic stuff in which are found the patterns—texts of all kinds, and experience of all kinds.

Now, we turn to finding patterns, which is to say how prediction is arrived at from the data that is given.

NOTES

1. Holmes (1881) op. cit. Prologue, p. xii, n. 9.
2. Id.
3. When writing for the Supreme Court on a question of the law of Puerto Rico, Justice Holmes reiterated his earlier idea about experience, here concluding that the judge *without* the experience ought to exercise restraint. The range of facts that Holmes identified as relevant are similar to those he identified forty years earlier in *The Common Law*:

   > This Court has stated many times the deference due to the understanding of the local courts upon matters of purely local concern... This is especially true in dealing with the decisions of a Court inheriting and brought up in a different system from that which prevails here. When we contemplate such a system from the outside it seems like a wall of stone, every part even with all the others, except so far as our own local education may lead us to see subordinations to which we are accustomed. But to one brought up within it, varying emphasis, tacit assumptions, unwritten practices, a thousand influences gained only from life, may give to the different parts wholly new values that logic and grammar never could have gotten from the books. Diaz et al. v. Gonzalez et al., 261 U.S. 102, 105–106, 43 S.Ct. 286, 287–88 (Holmes, J.) (1923).

   > Legal writers, in particular positivists, "have long debated which facts are the important ones in determining the existence and content of law." Barzun, 69 STAN. L. REV. 1323, 1329 (2017). Holmes's writings support a broad interpretation of "which facts..." he had in mind, and he was deliberate when he said that it is only "[t]he *theory* of our legal system... that the conclusions to be reached in a case will be induced only by evidence and argument in open court, and not by any outside influence": Patterson v. Colorado *ex rel.* Att'y Gen., 205 U.S. 454, 562 (1907) (emphasis added).

4. Gompers v. United States, 233 U.S. 604, 610 (1914).
5. See Rabban (2013) 215–68.
6. Chapter 5, pp. 54–57.
7. See further Chapter 10, pp. 114–119.
8. Kroll et al., op. cit., n. 76, at 660, point out that online learning systems pose additional challenges for algorithmic accountability.

CHAPTER 4

Finding Patterns as the Path from Input to Output

As [Judea Pearl] sees it, the state of the art in artificial intelligence today is merely a souped-up version of what machines could already do a generation ago: find hidden regularities in a large set of data. "All the impressive achievements of deep learning amount to just curve fitting," he said recently. [...]

The way you talk about curve fitting, it sounds like you're not very impressed with machine learning [remarks the interviewer]. "No, I'm very impressed, because we did not expect that so many problems could be solved by pure curve fitting. It turns out they can."

Judea Pearl as interviewed by Kevin Hartnett,
QUANTA MAGAZINE (May 15, 2018)[1]

Judea Pearl won the 2011 Turing Award, the "Nobel Prize for computer science," for his work on probabilistic and causal reasoning. He describes machine learning as "just curve fitting," the mechanical process of finding regularities in data. The term comes from draftsmen's use of spline curves, flexible strips made from thin pieces of wood or metal or plastic, to draw smooth lines through a set of pins.

In this chapter, we posit a further, specific analogy. We posit an analogy between Pearl's description of machine learning and Holmes's view of law. According to Holmes, the main task in law is finding patterns in human experience; law should not be seen simply as an exercise in mathematical logic. Likewise, machine learning should be thought of as curve fitting, i.e. as finding regularities in large datasets, and not as algorithms that execute a series of logical steps.

© The Author(s) 2020
T. D. Grant and D. J. Wischik, *On the path to AI*,
https://doi.org/10.1007/978-3-030-43582-0_4

We described in Chapter 2 why it is not helpful to view machine learning as an algorithm. It is not an adequate explanation of what makes machine learning the powerful tool it has become today to say that it is about executing a series of logical instructions, composed in a piece of programming code. To understand what makes machine learning distinctive one has to start with the role of datasets as input, a role we described in Chapter 3 above, and which may be analogized to Holmes's view of the jurist's experience. In this chapter we now examine pattern finding more closely, first in law then in machine learning, to see how far the analogy might go.

4.1   Pattern Finding in Law

Holmes said in *The Path of the Law* that identifying the law means "follow[ing] the existing body of dogma into its highest generalizations."[2] Two years after *The Path*, Holmes described law as a proposition that emerges when certain "ideals of society have been strong enough to reach that final form of expression."[3] To describe the law as Holmes did is to call for "the scientific study of the morphology and transformation of human ideas in the law."[4] If the pattern is strong enough, then the proposition emerges, the shape becomes clear.

Holmes returned a number of times to this idea that law is to be identified in patterns in human nature and practice. In a Supreme Court judgment in 1904, he addressed the right of "title by prescription." Under that right, a sustained and uncontested occupation of land can override a legal title to that land. Prescription is thus an example where the law explicitly recognizes that a pattern of reality on the ground *is* the law. Holmes described prescription like this:

> Property is protected because such protection answers a demand of human nature, and therefore takes the place of a fight. But that demand is not founded more certainly by creation or discovery than it is by the lapse of time, which gradually shapes the mind to expect and demand the continuance of what it actually and long has enjoyed, even if without right, and dissociates it from a like demand of even a right which long has been denied.[5]

This way of describing title by prescription evoked the search for pattern in experience. How society actually behaves and how people think about that behavior are facts in which a pattern may be discerned. If the pattern is well enough engrained, if it "shapes the mind" to a sufficient degree, and one knows how to discern it, then legal conclusions follow.

In what is perhaps his most famous dissenting opinion, that in *Lochner v. New York*, Holmes applied this idea about the ideals of society and the shape of the law a good deal further. The Supreme Court concluded that a New York state statute limiting the hours employees worked in a bakery violated the freedom of contract as embodied in the 14th Amendment. Holmes, as against his colleagues' formal reading of the 14th Amendment, argued that one should interpret the constitutional right in the light of the patterns of belief discernible in society:

> Every opinion tends to become a law. I think that the word 'liberty' in the 14th Amendment, is perverted when it is held to prevent the natural outcome of a dominant opinion, unless it can be said that a rational and fair man necessarily would admit that the statute proposed would infringe fundamental principles as they have been understood by the traditions of our people and our law.[6]

In the land title case, the rule of title by prescription acknowledged that the pattern in human experience *is* the law. A formal rule, exceptionally, there corresponded to what Holmes thought law is. In *Lochner*, by contrast, there was no formal rule that says you are to interpret the 14th Amendment by reference to "dominant opinion." The reading that Holmes arrived at in *Lochner* thus illustrates just how far-reaching Holmes's conception of the law as a process of pattern finding was. Even the plain text of the law, which a logician might think speaks for itself, Holmes said calls for historical analysis. The meaning of a text is not to be found only in its words, but in the body of tradition and opinion around it: "A word [in the Constitution] is not a crystal, transparent and unchanged, but the skin of a living thought."[7] Holmes believed that we identify the law by systematically examining the shape of what exists already and what might later come—"the morphology and transformation of human ideas."

A good jurist reaches decisions by discerning patterns of tradition and practice. The bad jurist treats cases as exercises in logical deduction. According to Holmes, "a page of history is worth a volume of logic."[8]

4.2   SO MANY PROBLEMS CAN BE SOLVED
BY PURE CURVE FITTING

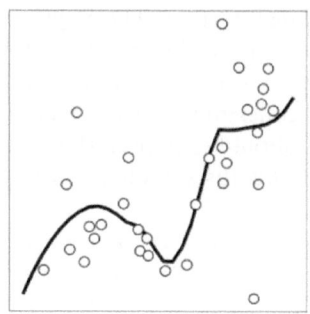

Judea Pearl expressed surprise that so many problems could be solved by curve fitting. And to someone from outside machine learning, it may seem preposterous that Holmes's pattern finding might be analogous to drawing a line through a collection of points, as illustrated in the figure above. To give some idea of the scope of what machine learning researchers express as curve fitting, we now consider some applications. We have chosen applications from law and data to keep with our analogy to legal pattern finding—but curve fitting applications from any number of application areas, such as those from the data science careers website that we listed in Chapter 1, would support the same point.

Our first application relates to Holmes's famous epigram "The prophecies of what the courts will do in fact, and nothing more pretentious, are what I mean by the law."[9] Suppose it were possible to draw a chart summarizing the body of relevant case law. Each case would be assigned an x coordinate encoding the characteristics of the case (the type of plea, the set of evidence, the history of the judge, and so on) and a y coordinate encoding the outcome of the case (the decision reached, the sentence, and so on), and a point would be plotted for each case at its assigned x and y coordinates. We could then draw a smooth curve that expresses how the y coordinate varies as a function of the x coordinate—i.e. find the pattern in the dataset—and we could use this curve to predict the likely outcome of a new case given its x coordinate.

This may sound preposterous, a law school version of plotting poets on a chalk board like the English teacher in *Dead Poets Society* did to ridicule a certain kind of pedantry.[10] However, it is an accurate description of how machines are able to accomplish such tasks as translating text or captioning images. A chalk board has only two dimensions; a machine learning system works in many more dimensions, represented through mathematical functions. The coordinates are expressed in sophisticated geometrical spaces (instead of x, use $x_1, x_2, \ldots, x_n$ for some large number of dimensions n) that go beyond human visualization abilities; but the method is nothing more than high dimensional curve fitting.

The above application is a thought experiment. Here are some actual examples borrowed from a recent book on LAW As DATA[11]:

(i) Predicting whether a bill receives floor action in the legislature, given the party affiliation of the sponsor and other features, as well as keywords in the bill itself.

(ii) Predicting the outcome of a parole hearing, given keywords that the inmate uses.

(iii) Predicting the case-ending event (dismissal, summary judgement, trial, etc.), given features of the lawsuit such as claim type or plaintiff race or plaintiff attorney's dismissal rate.

(iv) Predicting the topic of a case (crime, civil rights, etc.) given the text of an opinion. (To a human with a modicum of legal training this is laughably simple, but for machine learning it is a great achievement to turn a piece of text into a numerical vector $(x_1, x_2, \ldots, x_n)$ that can be used as the x coordinate for curve fitting. The mathematics is called "doc2vec".)

(v) Predicting the decision of an asylum court judge given features of the case. (If a prediction can be made based on features revealed in the early stages of a case, and if the prediction does not improve when later features are included, then perhaps the judge was sleeping through the later stages.)

We have used the word "predict" for all of these examples. Most of these tasks are predictive in the sense of forecasting, but in the case (iv) the word "predict" might strike a layperson as odd. In machine learning, the word "predict" is used even when the outcome being predicted is already known; what matters is that the outcome is not known *to the machine making the prediction*. Philosophers use the words "postdiction" or "retrodiction" for such cases. In Chapter 5 we address in detail why computer scientists use the language of prediction to describe the outputs of a machine learning system—and why Holmes used it to describe the outputs of law.

4.3   Noisy Data, Contested Patterns

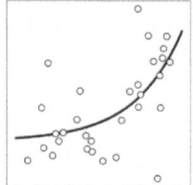 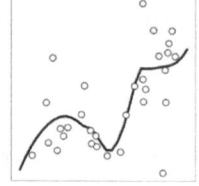 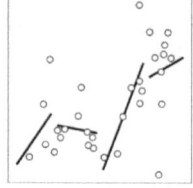 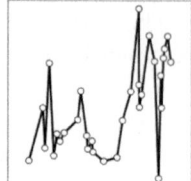

Holmes's wrote that "a page of history is worth a volume of logic." When lawmakers ask for "the logic involved"[12] in automated decision making, they should really be asking for "a story about the training dataset." It is the data—that which is a given and thus came before—that matters in machine learning, just as the history is what matters in Holmes's idea of law—not some formal process of logic.

But history can be contested. Even when parties agree on the facts, there may be multiple narratives that can be fitted.[13] Likewise, for a given dataset, there may be various curves that may be fitted, as the figure above illustrates. We might wish to remove the subjectivity, leaving us with a volume of irrefutable logic proving that the decision follows necessarily from the premises, but that is the nature neither of law nor of machine learning. The phrase "story about the training dataset" is meant to remind us of this.

For some datasets, there may be a clear curve that fits all the data points very closely. In Holmes's language, this corresponds to finding patterns in experience that have attained the "final form of expression." The process of finding the law, as Holmes saw it, is the process of finding a pattern strong enough to support such "highest generalizations." Not all "existing dogma" lends itself, however, to ready description as law; one does not always locate in the body of experience a "crystal, transparent." Likewise, not all datasets have a well-fitting curve; the y coordinates may simply be too noisy.

Some writers refer to machine learning systems as "inferring rules from data," "deriving rules from data," and the like.[14] We recommend the phrase "finding patterns in data," because it is better here to avoid any suggestion of clean precise law-like rules. The patterns found by machine learning are not laws of nature like Newton's laws of motion, and they are not precise stipulative rules in the sense of directives laid down in statutes. They are simply fitted curves; and if the data is noisy then the curves will not fit well.

While we have noted here that pattern finding is an element shared by machine learning and law, we should also note a difference. Law as Holmes saw it, and as it must be seen regardless of one's legal philosophy, is an activity carried out by human beings. Law involves intelligence and thought. Machine learning is not thought. Once the human programmer has decided which class of curves to fit, the machine "learning" process is nothing more than a mechanical method for finding the best fitting curve within this class. Caution about anthropomorphizing machine learning is timely because there is so much of it, not just in popular culture, but in technical writing as well—and it obscures what machine learning really is. Machine learning is not thought. It is not intelligence. It is not brain activity. Pearl described it as curve fitting to emphasize this point, to make clear it is nothing more than a modern incarnation of the draftsman's spline curve. That description does not entail any modesty at all about what machine learning can do. It only serves to illustrate how it is that machine learning does it.

Notes

1. Reprinted in The Atlantic (May 19, 2018).
2. 10 Harv. L. Rev. at 476.
3. Holmes, *Law in Science and Science in Law*, 12 Harv. L. Rev. 443, 444 (1898–1899).
4. Id. at 445.
5. Davis v. Mills, 194 U.S. 451, 457, 24 S.Ct. 692, 695 (Holmes, J., 1904).
6. *Lochner v. New York*, 198 U.S. 45, 75–76, 25 S.Ct. 539, 547 (Holmes, J., dissenting, 1905).
7. Towne v. Eisner, 245 U.S. 418, 425 (1918).
8. New York Trust Co. v. Eisner, 256 U.S. 345, 349 (1921).
9. 10 Harv. L. Rev. at 461.
10. The idea, which in the movie is read by a student from a fictive text entitled Understanding Poetry as the teacher draws a chart representing Byron and Shakespeare, is this:

If the poem's score for perfection is plotted along the horizontal of a graph, and its importance is plotted on the vertical, then calculating the total area of the poem yields the measure of its greatness.

Dead Poets Society (release date: June 9, 1989): http://www.dailyscript. com/scripts/dead_poets_final.html. For a critique, see Kevin J.H. Dettmar, *Dead Poets Society Is a Terrible Defense of the Humanities*, THE ATLANTIC (Feb. 19, 2014).

11. Taken from Livermore & Rockmore (eds.), LAW AS DATA. COMPUTATION, TEXT & THE FUTURE OF LEGAL ANALYSIS (2019). The examples are from Vlad Eidelman, Anastassia Kornilova & Daniel Argyle, *Predicting Legislative Floor Action* (id. 117–50); Hannah Laqueur & Anna Venancio, *A Computational Analysis of California Parole Suitability Hearings* (id. 191–230); Charlotte S. Alexander, Khalifeh al Jadda, Mohammad Javad Feizhollahi & Anne M. Tucker, *Using Text Analytics to Predict Litigation Outcomes* (id. 271–308); Elliott Ash & Daniel Chen, *Case Vectors: Spatial Representations of the Law Using Document Embeddings* (id. 309–34); Daniel Chen, *Machine Learning and the Rule of Law* (id. 429–38). The latter application is a thought experiment, and the first two are implemented case studies.

12. As at GDPR Arts. 13(2)(f), 14(2)(g), and 15(1)(h).

13. For examples of how jurists have applied Holmes's aphorism in disputable ways, see Bosmajian, 38 J. CHURCH STATE 397–409 (1996).

14. See for example Kroll et al. at 638.

CHAPTER 5

Output as Prophecy

In the preceding chapter we considered some of the purposes to which machine learning might be put—for example, to predict the topic of a court case given words that the judge used in the court's written judgment—and we described pattern finding as the method behind prediction. More important than the method however is the goal, in this example "to predict the topic," and in particular the keyword *predict*. We introduced that word in the preceding chapter to begin to draw attention to how computer scientists use it when they engineer machine learning systems. In machine learning systems, predictive accuracy is the be-all and end-all—the way to formulate questions, the basis of learning algorithms, and the metric by which the systems are judged. In this chapter we consider prediction, both in Holmes's view of law and in the machine learning approach to computing.

In Holmes's view of law, prediction is central. His answer to the question, *What constitutes the law?* has become one of the most famous epigrams in all of law:

> The prophecies of what the courts will do in fact, and nothing more pretentious, are what I mean by the law.[1]

Holmes's interest in the logic and philosophy of probability and statistics has come more to light thanks to recent scholarship[2]; he immersed

© The Author(s) 2020
T. D. Grant and D. J. Wischik, *On the path to AI*,
https://doi.org/10.1007/978-3-030-43582-0_5

himself early in his career in those subjects. Holmes's use of the word "prophecy" was deliberate. It accorded with his overall view of law by getting away from the scientific and rational overtones of "prediction," even as he used that word too. Arguably, given how elusive explanations have been of how machine learning systems arrive at the predictions they make, "prophecy" is a good term in that context too.

We expand in this chapter on Holmes's idea that prophecies constitute the law, and then we return to prediction in machine learning.

5.1   Prophecies Are What Law Is

Holmes's famous epigram has been widely repeated, but it is not widely understood. Taken in isolation from *The Path of the Law*, where Holmes set it down, and in isolation from Holmes's development as a thinker, it might sound like no more than a piece of pragmatic advice to a practicing lawyer: don't get carried away by the cleverness of your syllogisms; ask yourself, instead, what the judge is going to do in your client's case. If that is all it meant, then it would be good advice, but it would not be a concept of law. Holmes had in mind a concept of law. The epigram needs to be read in context:

> The confusion with which I am dealing besets confessedly legal conceptions. Take the fundamental question, What constitutes the law? You will find some text writers telling you that it is something different from what is decided by the courts of Massachusetts or England, that it is a system of reason, that it is a deduction from principles of ethics or admitted axioms or what not, which may or may not coincide with the decisions. But if we take the view of our friend the bad man we shall find that he does not care two straws for the axioms or deductions, but that he does want to know what the Massachusetts or English courts are likely to do in fact. I am much of this mind. The prophecies of what the courts will do in fact, and nothing more pretentious, are what I mean by the law.[3]

Holmes was contrasting "law as prophecy" to "law as axioms and deductions." He saw an inductive approach to law—the pattern finding approach that starts with data or experience—not just to improve upon or augment legal formalism. He saw it as a corrective. The declaration in his dissent in the *Lochner* case a few years after *The Path of the Law* that "[g]eneral propositions do not decide concrete cases"[4] was not just to say that the formal, deductive approach is insufficient; it was to say that formalism gets in the way.

The centrality of concern for Holmes was the reality of decision, the output that a court might produce. The realism or positivism in this understanding of law contrasted with the formalist school that had long prevailed. To shift the concern of lawyers in this way was to lessen the role of doctrine, of formal rules, and to open a vista of social and historical considerations heretofore not part of the law school curriculum and ignored, or at any rate not publicly acknowledged, by lawyers or judges. Jurists have been divided ever since as to whether the shift of conception was for better or worse. Whatever one's assessment of it, the concept of law as Holmes expressed it continues to influence the law.

There is more still to Holmes's epigram about prophecies. True, the contrast it entails between the inductive method and the deductive method alone has revolutionary implications. But Holmes was not merely concerned with what method "our friend the bad man" (or indeed the bad man's lawyer) should employ to predict the outcome of a case. He wasn't writing a law practice handbook. He was interested in individual encounters with the law to be sure,[5] but this was because he sought to reach a general understanding of law *as a system*. Holmes's invocation of prophecies, like his use of terms from logic and mathematics, was memorable use of language, but it was more than rhetoric: it was at the core of Holmes's definition of law. He referred to the law as *"systematized prediction."*[6] This was to apply the term "prediction" broadly—indeed across the legal system as a whole. Holmes was not sparing in his use of the word "prophecy" when defining the law. The word "prophesy" or its derivates appear nine times in *The Path of the Law*.[7] He used it in the same sense when writing for the Supreme Court.[8] Holmes's concern with prediction is traceable in his other writings too.[9] The heart of Holmes's insight, and what has so affected jurisprudence since, is that the law *is* prediction.[10] Prophecy does not refer solely to the method for predicting what courts will do. Prophecy is what *constitutes* the law.

Prophecy of what, by whom, and on the basis of which input data?

Holmes gave several illustrations. For example, he famously described the law of contract as revolving around prediction: "The duty to keep a contract at common law means a prediction that you must pay damages if you do not keep it, and nothing else."[11] He stated his main thesis in similar terms: "a legal duty so called is nothing but a prediction that if a man does or omits certain things he will be made to suffer in this or that way by judgment of the court."[12] This is a statement about "legal duty" irrespective of the content of the duty. It thus just as well describes any duty that exists in the legal system.

We think that Holmes's concept of law as prediction indeed is comprehensive. Many jurists don't see it that way. Considering how Holmes understood law to relate to decisions taken by courts, one sees why his concept of law-as-prophecy often has received a more limited interpretation.

Holmes wrote that "the object of [studying law], then, is prediction, the prediction of the incidence of the public force through the instrumentality of the courts."[13] Making the equation directly, he wrote, "Law is a statement of the circumstances, in which the public force will be brought to bear upon men through the courts....."; a "word commonly confined to such prophecies... addressed to persons living within the power of the courts."[14] It is often assumed that Holmes's description here does not account for the decisions of the highest courts in a jurisdiction, courts whose decisions are final. After all, in a system of hierarchy, the organ at the apex expects its commands to be obeyed. To call decisions that emanate from such quarters "predictions" seems to ignore the reality of how a court system works. In a well-functioning legal system, a judgment by a final court of appeal saying, for example, that the police are to release such and such a prisoner, should lead almost certainly to that outcome. The court commands it; the prisoner is released.

In two respects, however, one perhaps trivial but the other assuredly significant, the highest court's statements, too, belong to the concept of law as prophecy.

First, even in a well-functioning legal system, the court's decision is still only a prediction. As outlandish as the situation would be in which the police ignored the highest court, it is a physical possibility. A statistician might say that the probability is very high (say, 99.9999%) that the highest court's judgment that the law requires the prisoner to be released will in fact result in an exercise of public power in accordance with that judgment. We will say more below about the relation between probability and prediction.[15] Leaving that relation aside for the moment, a judgment even of the highest court is a prediction in Holmes's sense. It is a prediction in this way: the implementation of a judgment by agents of public power is an act of translation, and in that act the possibility exists for greater or lesser divergence from the best understanding of what the judge commanded. So the definition of law as prophecy is instanced in the chance that the "public force" will not properly implement the judicial decision. In a well-functioning legal system, the chance is remote. In legal systems that don't function well, the predictive character of final judgments is more immediate, because the risk in those systems is greater that the public force will not properly implement the courts' commands. "Finality" in some judicial systems is more formal than real.[16]

The further respect in which the concept of law as prophecy is comprehensive comes to view when we consider how judges decide cases and how advocates argue them. In deciding a case, a judge will have in mind how that decision is likely to be interpreted, relied upon, or rejected, by future courts and academics and public opinion, as well as by the instruments of public force. The barrister, for her part, in deciding what line of argument to pursue, will have in mind how the judge might be influenced, and this in turn requires consideration of the judge's predictions about posterity. Holmes thus described a case after the highest court had decided it as still in the "early stages of law."[17] As Kellogg puts it, Holmes situated a case "not according to its place on the docket but rather in the continuum of inquiry into a broader problem."[18]

The law is a self-referential system, whose rules and norms are consequences of predictions of what those rules and norms might be.[19] Some people participate in the system in a basic and episodic way, for example the "bad man" who simply wants advice from a solicitor about likely outcomes in respect of his situation. Some people participate in a formative way. The apex example is the judge of the final court of appeal whose predictions about the future of the legal system are embodied in a judgment which she expects as a consequence of her authority in the legal system to be a perfect prediction of the exercise of public power in respect of the case. But her outlook, indeed her self-regard as a judge, entails more than that; a judge does more than participate in disconnected episodes of judging: she hopes that any judgment she gives in a case, because she strives for judgments that withstand the test of time, will be a more or less accurate prediction of how a future case bearing more or less likeness to the case will be decided. The judge describes her judgment as command, not as prophecy; but the process leading to it, and the process as the judge hopes it will unfold in the future, is predictive. Law-as-prophecy, understood this way, has no gap.

Holmes's claim, as we understand it, holds law to be predictive through and through. Prophecy is what law is made of. The predictive character of law, in this constitutive sense, is visible in the process of judicial decision, regardless what level of the judiciary is deciding; and it is visible in all other forms of legal assertion as well. Prophecy embraces all parts of legal process.

So everyone who touches the law is making predictions, from the self-interested "bad man" to the judge in the highest court of the land, and they make predictions about the full range of possible outcomes. The experience that influences their predictions, as we saw in Chapter 3,[20] Holmes understood to be wide, and the new situations that they make predictions about are unlimited. People on Holmes's path of the law thus engage in tasks much broader than standard machine learning tasks. As we also discussed in Chapter 3,[21] machine inputs, while they consist in very large datasets ("Big Data"), are limited to inputs that have been imparted a considerable degree of structure—a degree of structure almost certainly lacking in the wider (and wilder) environment from which experience might be drawn. Machine outputs are correspondingly limited as well. Let us now further explore the machine learning side of the analogy—and its limits.

5.2   PREDICTION IS WHAT MACHINE LEARNING OUTPUT IS

Holmes, writing in 1897, obviously did not have machine learning in mind. Nevertheless, his idea that prophecy constitutes the law has remarkable resonance with machine learning, a mechanism of computing that, like law as Holmes understood it, is best understood as constituted by prediction.

The word *prediction* is a term of art in machine learning. It is used like this:

> In a typical scenario, we have an outcome measurement, usually quantitative (such as a stock price) or categorical (such as heart attack/no heart attack), that we wish to predict based on a set of features (such as diet and clinical measurements). We have a training set of data, in which we observe the outcome and feature measurements for a set of objects (such as people). Using this data we build a prediction model, or learner, which will enable us to predict the outcome for new unseen objects. A good learner is one that accurately predicts such an outcome.[22]

Though in common parlance the term "prediction" means forecasts—that is to say, statements about future events—in machine learning the term has a wider meaning. We have touched on the wider meaning in Chapter 4 and at the opening of the present chapter. Let us delve a little more into that wider meaning now.

It is true that some machine learning outputs are "prediction" in the sense in which laypersons typically speak: "Storm Oliver will make landfall in North Carolina"[23] or "the stock price will rise 10% within six months." Other outputs are not predictions in the layperson's sense. Indeed, the main purposes for which machine learning is used do not involve predictions of that kind—purposes like classifying court cases by topic or controlling an autonomous vehicle. Whatever the purposes for which it is used, machine learning involves "prediction" of the more general kind computer scientists denote with that term.

The essential feature of prediction in machine learning is that it should concern "the outcome for new unseen objects," i.e. for objects not in the training set. Thus, for example, if the training set consists of labelled photographs, and if we treat the pixels of the photograph as features and the label as the outcome, then it is prediction when the machine learning system is given a new photograph as input data and it outputs the label "kitten." In machine learning prediction, "pre-" simply refers to *before the true outcome measurement has been revealed to the machine learning system*. The sense of "pre-" in "prediction" holds even though other parties might well already know the outcome. For example, the computer scientist might well already know that the new photograph is of a tiger, not a kitten. That assignment of label-to-picture has already happened, but the machine learning system has not been told about it at the point in time when the system is asked to predict. Philosophers of science use the terms "postdiction" or "retrodiction" to refer to predicting things that have already happened.[24] These words are not used in the machine learning community, but the concept behind them is much what that community has in mind when it talks about prediction.

A significant part of the craft of machine learning is to formulate a task as a prediction problem. We have already described how labelling a photograph can be described as prediction. A great many other examples may be given. Translation can be cast as prediction: "predict the French version of a sentence, given the English text," where the training set is a human-translated corpus of sentences. Handwriting synthesis can as well. Given a dataset of handwritten text, recorded as the movements of a pen nib, and given the same text transcribed into text in a word processor, the task of handwriting synthesis can be cast as prediction: "predict the movements of a pen nib, given text from a word processor." As Judea Pearl observed in the interview with which we opened Chapter 4,[25] it is truly remarkable how many tasks can be formulated this way. In the social sciences, it is "a rather new epistemological approach [...] and research agendas based on predictive inference are just starting to emerge."[26] A theory of law based on predictive inference, however, emerged over a century ago: Holmes theorized law to be constituted by prophecy. So too might we say that machine learning is constituted by prediction.

Moreover, prediction is not just the way that machine learning tasks are formulated. It is also the benchmark by which we train and evaluate machine learning systems in the performance of their tasks. The goal of training is to produce a "good learner," i.e. a system that makes accurate predictions. Training is achieved by measuring the difference between the machine's predictions (or postdictions, as the philosophers would say) and the actual outcomes in the training dataset; and iteratively tweaking the machine's parameter values so as to minimize the difference. The machine that reliably labels tigers as "tigers" has learned well and, at least for that modest task, needs no more tweaking. The machine that labels a tiger as a "kitten" needs tweaking. The one that labels a tiger as "the forests of the night," though laudable if its task had been to predict settings in which tigers are found in the poetry of William Blake, needs some further tweaking still to perform the task of labeling animals. This process of iterative tweaking, as we noted in Chapter 2, is what is known as gradient descent,[27] the backbone of modern machine learning. Thus, a mechanism of induction, not algorithmic logic, is at the heart of machine learning, much as Holmes's "inductive turn" is at the heart of his revolutionary idea of law.

It is not machine learning's fundamental characteristic that it can be used to forecast future events—when will the next hurricane occur, where will it make landfall? One doesn't need machine learning to make forecasts. One can make forecasts about hurricanes and the like with dice or by sacrificing a sheep (or by consulting a flock of dyspeptic parrots). One can also make such forecasts with classic algorithms, by simulating dynamical systems derived from atmospheric science. This sort of prediction is not the fundamental characteristic of machine learning.

The fundamental characteristic of machine learning is that the system is *trained using a dataset* consisting of examples of input features and outcome measurements; until, through the process of gradient descent, the machine's parameter values are so refined that the machine's predictions, when we give it further inputs, differ only minimally from the actual outcomes in the training dataset. Judges, litigants, and their lawyers certainly try to align their predictive statements of law with what they discern to be the relevant pattern in law's input data, that is to say in the collective experience that shapes the law. It is equally the case, in Holmes's understanding of the law, that we do not test court judgments by comparing against stipulated "correct" labels the way our spam email or tiger detector was tested. Judgments are, however, tested against future judgments. This is to the point we made earlier about the judge's aim that her judgments withstand the test of time. The test is whether future judgments show her judgment to have been an accurate prediction, or at least not so far off as to be set aside and forgotten.

A machine learning system must be trained on a dataset of input features and outcome measurements. This is in contrast to the classic algorithmic approach, which starts instead from rules. For example the classic approach to forecasting the weather works by solving equations that describe how the atmosphere and oceans behave; it is based on scientific laws (which are presumably the result of codifying data from earlier experiments and observation). Just as machine learning rejects rules and starts instead with training data, Holmes rejected the idea that law is deriving outcomes based on general principles, and he cast it instead as a prediction problem—prophesying what a court will do—to be performed on the basis of experience.

5.3   LIMITS OF THE ANALOGY

As we noted in Chapter 3,[28] the predictions made by a machine learning system must have the same form as the outcomes in the training dataset, and the input data for the object to be predicted must have the same form as objects already seen. In earlier applications of machine learning, "same form" was very narrowly construed: for example, the training set for the ImageNet challenge[29] consists of images paired with labels; the machine learning task is to predict which one of these previously seen labels is the best fit for a new image, and the new image is required to be the same dimensions as all the examples in the training set. Human ability to make predictions about novel situations is far ahead of that of machines. A human lawyer can extrapolate from experience and make predictions about new cases that don't conform to a narrow definition of "cases similar to those already seen." The distance is closing, however, as researchers develop techniques to broaden the meaning of "same form." For example, an image captioning system[30] is now able to generate descriptions of images, rather than just repeat labels it has already seen. Thus, it is well within their grasp for machines to label an image as "tiger on fire in a forest," but they are still a long way, probably, from describing, as the poet did, the tiger's "fearful symmetry."

There is a more significant difference between predictions in machine learning and in law. In machine learning, the paradigm is that there is something for the learning agent—i.e., the machine—to learn. The thing to learn is data, something that is given, not a changing environment affected by numerous factors—including by the learning agent. A machine for translating English to French can be trained using a human-translated corpus of texts, and its translations can be evaluated by how well they match the human translation. Whatever translations the machine comes up with they do not alter the English nor French languages. In law, by contrast, the judgment in a case becomes part of the body of experience to be used in subsequent cases. Herein, we think, Holmes's concept of law as a system constituted from prediction may hold lessons for machine learning. In Chapters 6–8, we will consider some challenges that machine learning faces, and possible lessons from Holmes, as we discuss "explainability" of machine learning outputs[31] and outputs that may have invidious effects because they reflect patterns that emerge from the data (such as patterns of racial or gender discrimination).[32] In Chapter 9,[33] we will suggest that Holmes, because he understood law to be a self-referential process in which each new prediction shapes future predictions, might point the way for future advances in machine learning.

Before we get to the challenges of machine learning and possible lessons for the future from Holmes, we will briefly consider a question that prediction raises: does prediction, whether as the constitutive element of law or as the output of machine learning, necessarily involve the assessment of probabilities?

5.4   PROBABILISTIC REASONING AND PREDICTION

"For the rational study of the law the blackletter man may be the man of the present, but the man of the future is the man of statistics," said Holmes.[34] It is not certain that Holmes thought that the predictive character of law necessarily entails a probabilistic character for law. He was certainly interested in probability. In the time after his Civil War service, a period that Frederic Kellogg closely examined in Oliver Wendell Holmes, Jr. AND LEGAL LOGIC, Holmes studied theories of probability and was much engaged in discussions about the phenomenon, including how it relates to logic and syllogism.[35] Later, as a judge, he recognized the part played by probability in commercial life, for example in the functioning of the futures markets.[36] In personal correspondence, Holmes said that early in his life he had learned "that I must not say *necessary* about the universe, that we don't know whether anything is necessary or not. So that I describe myself as a *bet*tabilitarian. I believe that we can *bet* on the behavior of the universe..."[37] Holmes would have been comfortable with the idea that law, in its character as prediction, concerned probability as well. Some jurists indeed have discerned in Holmes's idea of law-as-prophecy just such a link.[38]

Predictions made by machine learning are not inherently probabilistic. For example, the "*k* nearest neighbors"[39] machine learning algorithm is simply "To predict the outcome for a new case, find the *k* most similar cases in the dataset, find their average outcome, and report this as the prediction." The system predicts a value, which may or may not turn out to be correct. Modern machine learning systems such as neural networks, however, are typically designed to generate predictions using the language of probability, for example "the probability that this given input image depicts a kitten is 93%."[40]

Separately, we can classify machine learning systems by whether or not they employ probabilistic *reasoning* to generate their predictions:

> [One type of] Machine Learning seeks to learn [probabilistic] models of data: define a space of possible models, learn the parameters and structure of the models from data; make predictions and decisions. [The other type of] Machine Learning is a toolbox of methods for processing data: feed the data into one of many possible methods; choose methods that have good theoretical or empirical performance; make predictions and decisions.[41]

Are legal predictions expressed in the language of probability? Lawyers serving clients do not always give probability assessments when they give predictions, but sometimes they do.[42] Some clients need such an assessment for purposes of internal controls, financial reporting, and the like. Others ask for it for help in strategizing around legal risk. Modern empirical turns in law scholarship, it may be added, are much concerned with statistics.[43] Attaching a probability to a prediction of a legal outcome is an inexact exercise, but it is not unfamiliar to lawyers.

Holmes, when he referred to the prophecies of what courts will do, is often read to mean that the law should be made readily predictable.[44] Though we don't doubt he preferred stable judges to erratic ones, we don't see that that was Holmes's point. Courts whose decisions are hard to predict are no less sources of legal decision. Even when the lawyer has the privilege to argue in front of a "good" judge, whom for present purposes we define as a judge whose decisions are easy to predict, the closer the legal question, the harder it is to predict the answer. It is inherent that lawyers will be more confident in some of their predictions than in others.

Judges, practically by definition of their role as legal authorities, do not proffer a view as to the chances that their judgments are correct. It is hard to see how the process of judgment would keep the confidence of society, if every judgment were issued with a p-value![45] Yet reading judgments through a realist's glasses, one may discern indicia of how likely it is that the judgment will be understood in the future to have stated the law. Judges do not shy from describing some cases as clear ones; others as close ones. They don't call it hedging, but that's very much what it's like. When a judge refers to how finely balanced such and such a question was, it has the effect of qualifying the judgment. It thus may be that one can infer from a judgment's text how much confidence one should have in the judgment as a prediction of future results. The text, even where it does not express anything in terms about the closeness of a case, still may give clues. The structure of the reasoning may be a clue: the more complex and particularistic a judge's reasoning, the more the judgment might be questioned, or at least limited in its future application. Textual clues permit an inference as to how confident one should be that the judgment accurately reflects a pattern in the experience that was the input behind it.[46]

Does the law use probabilistic reasoning to arrive at a prediction? In other words, once a judgment has been made and it becomes part of the body of legal experience, do lawyers and judges reason about their *level of confidence* that an earlier judgment is relevant for their predictions about a current case? Ex post, every judgment is in fact, to a greater or lesser extent, questioned or rejected or ignored—or affirmed or relied upon. Nullification, reversal, striking down—by whatever term the legal system refers to the process, a rejection of a judgment by a controlling authority is a formal expression that the judge got it wrong.[47] Endorsement, too, is sometimes formal and explicit, the archetype being a decision on appeal that affirms the judgment. Formal and explicit signals, whether of rejection or of reliance, entail a significant adjustment in how much confidence we should have in a judgment as a prediction of a future case.

It is not just in appeals that we look for signals as to how confident we should be in a given judgment as a prediction of future cases. Rejection or endorsement might occur in a different case on different facts (i.e., not on appeal in the same case) and in that situation is therefore only an approximation: ignoring, rejecting, or "distinguishing" a past judgment; or invoking it with approval, a judge in a different case says or implies that the judge in the past judgment had the law wrong or he had it right, but indirect treatment in the new judgment, whether expressed or implied, says only so much about the past one. A jurist, considering such indirect treatment, would struggle to arrive at a numerical value to adjust how much confidence to place in the past judgment.[48] In evidence about judgments—evidence inferable from the words of the judgments themselves and evidence contained in their reception—one nevertheless discerns at least rough markers of the probability that they will be followed in the future.

There is no received view as to what Holmes thought the function of probability is in prediction. As is the case with machine learning, jurists make probabilistic as well as non-probabilistic predictions. You can state the law—i.e., give a prediction about the future exercise of public power—without giving an assessment of your confidence that your prediction is right. Jurists also use both probabilistic and non-probabilistic *reasoning*. Holmes, when referring to prophecies, was not however telling courts how to reason (or for that matter, legislatures or juries; we will return to juries in Chapter 7). His concern was to state what it is that constitutes the law. True, we don't call wobbly or inarticulate judges good judges. But Holmes was explicitly *not* concerned with the behavior of the "good" litigant; and, in his thinking about the legal system as a whole, his concern was not limited to the behavior of the "good" judge.

NOTES

1. Holmes, *The Path of the Law*, 10 HARV. L. REV. 457, 461 (1896–1897).
2. See Kellogg (2018) op. cit.
3. 10 HARV. L. REV. at 460–61 (1896–1897).
4. *Lochner*, 198 U.S. 45, 76 (1905) (Holmes, J., dissenting).
5. Though legal advice was not what Holmes was giving, his writings supply ample material for that purpose, and so have judges sometimes read him: see, e.g., Parker v. Citimortgage, Inc. et al., 987 F.Supp.2d 1224, 1232 n 19 (2013, Jenkins, SDJ).
6. 10 HARV. L. REV. at 458 (emphasis added).
7. The text, which runs to 21 pages (less than 10,000 words), contains the word "prophecy," "prophecies," or the verb "to prophecy" on five pages: 10 HARV. L. REV. at 457, 458, 461, 463, and 475.
8. American Banana Company v. United Fruit Company, 213 U.S. 347, 357, 29 S.Ct. 511, 513 (Holmes, J., 1909).
9. Moskowitz emphasized this line of Holmes's thought in *The Prediction Theory of Law*, 39 TEMP. L.Q. 413, 413–16 (1965–1966).
10. 10 HARV. L. REV. at 462.
11. Id. at 462.
12. Id.
13. Id. at 457.
14. American Banana Company, 213 U.S. at 357, 29 S.Ct. at 513.
15. This chapter, pp. 59–61.
16. See for example White, *Putting Aside the Rule of Law Myth: Corruption and the Case for Juries in Emerging Democracies*, 43 CORN. INT'L L.J. 307, 321 n. 118 (2010) (reporting doubt whether Mongolia's other branches of government submit to judicial decisions, notwithstanding the Constitutional Court's formal power of review). The predictive character of judgments of interstate tribunals, in this sense, is pronounced. Judgments of the International Court of Justice, to take the main example, under Article 60 of the Court's Statute are "final and without appeal," but no executive apparatus is generally available for their enforcement, and the Court has no ancillary enforcement jurisdiction. Even in the best-functioning legal system, high courts may have an uneasy relation to the executive apparatus whose conduct they sit in judgment upon. Recall Justice Frankfurter's concurrence in Korematsu v. United States where, agreeing not to overturn wartime measures against persons of Japanese, German, and Italian ancestry, he declared "[t]hat is their [the Government's] business, not ours": 323 U.S. 214, 225, 65 S.Ct. 193, 198 (1944) (Frankfurter, J., concurring).
17. Vegelahn v. Guntner & others, 167 Mass. 92, 106 (1896) (Field, C.J. & Holmes, J., dissenting).

18. Kellogg (2018) at 82.
19. See also Kellogg at 92: "Prediction had a broader and longer-term reference for [Holmes] than immediate judicial conduct, and was connected with his conception of legal 'growth'." See further Chapter 9.
20. Chapter 3, pp. 34–35.
21. Chapter 3, p. 38.
22. Hastie et al. (2009) 1–2.
23. It is the use of machine learning to make "predictions" in this sense ("predictive analytics") on which legal writers addressing the topic to date largely have focused. See, e.g., Berman, 98 B.U. L. Rev. 1277 (2018).
24. The terms "postdiction" and "retrodiction" are sometimes used in legal scholarship too, though writers who use them are more likely to do so in connection with other disciplines. See, e.g., Guttel & Harel, *Uncertainty Revisited: Legal Prediction and Legal Postdiction*, 107 Mich. L. Rev. 467–99 (2008), who considered findings from psychology that people are less confident about their postdictions (e.g., what was the outcome of the dice roll that I just performed?) than their predictions (e.g., what will be the outcome of the dice roll that I am about to perform?) id. 471–79.
25. See Chapter 4, p. 41.
26. Dumas & Frankenreiter, *Text as Observational Data*, in Livermore & Rockmore (eds.) (2019) at 63–64.
27. As to gradient descent, see Chapter 2, p. 23.

> Gradient descent is often coupled with another technique called *cross validation*, also based on prediction. The term derives from the idea of a "validation dataset." When training a machine learning system, it is not possible to measure prediction accuracy by testing predictions on the same dataset as was used to train the machine. (This can be shown mathematically.) Therefore, the training dataset is split into two: one part for training parameter values, the other part for measuring prediction accuracy. This latter part is called the "validation dataset." Cross validation is totemic in machine learning: stats.stackexchange.com, a popular Internet Q&A site for machine learning, calls itself *CrossValidated*. It is also technically subtle. See Hastie, Tibshirani & Friedman (2009) in §7.10, for a formal description

28. Chapter 3, p. 38.
29. Russakovsky, Deng et al., op. cit. (2015).
30. Hossain, Sohel, Shiratuddin & Laga, ACM CSUR 51 (2019).
31. Chapter 6, p. 70.
32. Chapter 7, pp. 81–88; and Chapter 8, pp. 89–100.

33. Chapter 9, pp. 103–111.
34. 10 Harv. L. Rev. at 469.
35. Kellogg (2018) 36–53.
36. Board of Trade of the City of Chicago v. Christie Grain & Stock Company et al., 198 U.S. 236, 247, 25 S.Ct. 637, 638 (Holmes, J., 1905). See also Ithaca Trust Co. v. United States, 279 U.S. 151, 155, 49 S.Ct. 291, 292 (Holmes, J., 1929) (mortality tables employed to calculate for purposes of tax liability the value of a life bequest as of the date it was made).
37. Letter from Holmes to Pollock (Aug. 30, 1929), reprinted De Wolfe Howe (ed.) (1942) vol. 2, p. 252 (emphasis original). Kellogg quotes this passage: Kellogg (2018) at 52.
38. See for example Coastal Orthopaedic Institute, P.C. v. Bongiorno & anthr, 807 N.E.2d 187, 191 (2004, Appeals Court of Massachusetts, Bristol, Berry J.). Cf. observing a relation between uncertainty and the predictive character of law, Swanson et al. v. Powers et al., 937 F.2d 965, 968 (1991, 4th Cir. Wilkinson, CJ):

> The dockets of courts are testaments... to the many questions that remain reasonably debatable. Holmes touched on this uncertain process when he defined 'the law' as '[t]he prophecies of what the courts will do'.

39. Hastie et al., op. cit. (n. 27) § 2.3.2. This simple description of the k nearest neighbor algorithm does not reflect the real cleverness, which consists in inventing a useful similarity metric such that the simple algorithm produces good predictions. Even cleverer is to use a neural network to learn a useful similarity metric from patterns in the data.
40. Id., § 4.1.
41. Lecture given by Zoubin Ghahramani at MIT, 2012. http://mlg.eng.cam.ac.uk/zoubin/talks/mit12csail.pdf.
42. Lawyers indeed give probability assessments to their clients often enough that behavioral decision theorists have studied the factors that influence lawyers' views as to the chances of winning or losing in court. See, e.g., Craig R. Fox & Richard Birke, *Forecasting Trial Outcomes: Lawyers Assign Higher Probability to Possibilities That Are Described in Greater Detail,* 26(2) Law Hum. Behav. 159–73 (2002).
43. As to empiricism in legal scholarship generally, see Epstein, Friedman & Stone, *Foreword: Testing the Constitution,* 90 N.Y.U. L. Rev. 1001 and works cited id. at 1003 nn. 4, 5 and 1004 n. 6 (2015); in one of law's subdisciplines, Shaffer & Ginsburg, *The Empirical Turn in International Legal Scholarship,* 106 AJIL 1 (2012).
44. See for example Kern v. Levolor Lorentzen, Inc., 899 F.2d 772, 781–82 (1989, 9th Cir., Kozinski, C.J., dissenting).

45. I.e., a numerical value representing the probability that a future court will *not* treat the judgment as a correct statement of law. A *"p*-value" is a term familiar to courts, but not one they use to describe their own judgments. See Matrixx Initiatives, Inc. v. Siracusano, 563 U.S. 27, 39, 131 S.Ct. 1309, 1319 n. 6 (Sotomayor, J., 2011):

> 'A study that is statistically significant has results that are unlikely to be the result of random error....' To test for significance, a researcher develops a 'null hypothesis'—*e.g.*, the assertion that there is no relationship between Zicam use and anosmia... The researcher then calculates the probability of obtaining the observed data (or more extreme data) if the null hypothesis is true (called the *p*-value)... Small *p*-values are evidence that the null hypothesis is incorrect. (citations omitted)

See also In re Abilify (Aripiprazole) Products Liability Litigation, 299 F.Supp.3d 1291, 1314–15; Abdul-Baaqiy v. Federal National Mortgage Association (Sept. 27, 2018) p. 7.

46. On a court consisting of more than one judge and on which it is open to members of the court to adopt separate or dissenting opinions, the existence and content of such opinions are another source of evidence as to how much confidence one might place in the result. The possibility of assigning confidence intervals to judgments on such evidence is suggested here: Posner & Vermeule, *The Votes of Other Judges*, 105 Geo. L.J. 159, 177–82 (2016). Regarding the influence of concurring opinions on future judgments, see Bennett, Friedman, Martin & Navarro Smelcer, *Divide & Concur: Separate Opinions & Legal Change*, 103 Corn. L. Rev. 817 (2018), and in particular the data presented id. at 854 and *passim*. Cf. Eber, *Comment, When the Dissent Creates the Law: Cross-Cutting Majorities and the Prediction Model of Precedent*, 58 Emory L.J. 207 (2008); Williams, *Questioning Marks: Plurality Decisions and Precedential Constraint*, 69 Stan. L. Rev. 795 (2017); *Plurality Decisions— The Marks Rule—Fourth Circuit Declines to Apply Justice White's Concurrence in Powell v. Texas as Binding Precedent—Manning v. Caldwell*, 132 Harv. L. Rev. 1089 (2019).

47. Or got something in the judgment wrong while having gotten other things right. We speak above, for sake of economy of expression, about a judgment struck down *in toto*.

48. A recent study, though for different purposes, makes an observation apt to our point: "It is one thing to say that the standards of juridical proof are to be explicated in probabilistic terms, *it is another to provide such an explication*." Urbaniak (2018) 345 (emphasis added).

CHAPTER 6

Explanations of Machine Learning

The danger of which I speak is [...] the notion that a given system, ours, for instance, can be worked out like mathematics from some general axioms of conduct. [...] This mode of thinking is entirely natural. The training of lawyers is a training in logic. The processes of analogy, discrimination, and deduction are those in which they are most at home. The language of judicial decision is mainly the language of logic. And the logical method and form flatter that longing for certainty and for repose which is in every human mind. But certainty generally is illusion, and repose is not the destiny of man.

Oliver Wendell Holmes, Jr., The Path of the Law *(1897)*

Calls for "explainability" of machine learning outputs are much heard today, in academic and technical writing as well as in legislation such as the GDPR, the European Union's General Data Protection Regulation.[1] We are not convinced that very many lawmakers or regulators understand what would need to be done, if the explainability they call for is to be made meaningful.[2] It might seem to a legislator, accustomed to using the logical language of law and legal decisions, that an algorithmic decision "can be worked out like mathematics from some general axioms." In the law, Holmes rejected the idea that a logical argument gives a satisfactory explanation of a judicial decision. Instead, in a play upon Aristotle's system of logic, he invoked the "inarticulate major premise."[3] As a method to account for legal decision making, that idea, as Holmes employed it, left to one side the formal logic that jurists classically had employed to explain a legal output. In this chapter we will suggest that Holmes's idea

© The Author(s) 2020
T. D. Grant and D. J. Wischik, *On the path to AI*,
https://doi.org/10.1007/978-3-030-43582-0_6

of the inarticulate major premise offers a better way to think about explanations in machine learning—and also throws fresh light on a fundamental philosophical stance in machine learning, the "prediction culture."

6.1   Holmes's "Inarticulate Major Premise"

The premise behind a decision, it was Holmes's view, is not always expressed. Legal decision-makers offer an *apologia*, a logical justification for the decision they have reached, but the real explanation for a decision is to be found in the broad contours of experience that the decision-maker brings to bear. As Holmes put it in 1881 in *The Theory of Interpretation*, decision-makers "leave their major premises inarticulate."[4]

Holmes addressed this phenomenon again, and most famously, in his dissent in *Lochner v. New York*. To recall, the Supreme Court was asked · to consider whether a New York state law that regulated working hours in bakeries and similar establishments was constitutionally infirm. The majority decided that it was. According to the majority, the law interfered with "liberty" as protected by the 14th Amendment of the Constitution. Holmes in his dissent wrote as follows:

> Some of these laws embody convictions or prejudices which judges are likely to share. Some may not. But a Constitution is not intended to embody a particular economic theory, whether of paternalism and the organic relation of the citizen to the state or of laissez faire. It is made for people of fundamentally differing views, and the accident of our finding certain opinions natural or familiar, or novel, and even shocking, ought not to conclude our judgment upon the question whether statutes embodying them conflict with the Constitution of the United States.
>
> General propositions do not decide concrete cases. The decision will depend on a judgment or intuition more subtle than any articulate major premise. But I think that the proposition just stated, if it is accepted, will carry us far toward the end. Every opinion tends to become a law. I think that the word 'liberty' in the 14th Amendment, is perverted when it is held to prevent the natural outcome of a dominant opinion, unless it can be said that a rational and fair man necessarily would admit that the statute proposed would infringe fundamental principles as they have been understood by the traditions of our people and our law. It does not need research to show that no such sweeping condemnation can be passed upon the statute before us.[5]

Holmes rejected the straightforward logical deduction that would have read like this: *The 14th Amendment protects liberty; the proposed statute limits freedom of contract; therefore the proposed statute is unconstitutional.* He did not accept that this "general proposition[...]" contained in the 14th Amendment could "decide concrete cases," such as the case that the New York working hours law presented in *Lochner*. It was Holmes's suspicion that *laissez faire* economic belief on the part of the other judges was the inarticulate premise lurking behind the majority opinion, the premise that had led them to offer their particular deduction. Holmes posited instead that the meaning of the word "liberty" in the 14th Amendment should be interpreted in the light of the "traditions of our people and our law" and that applying "judgement or intuition" to the state of affairs prevailing in early twentieth century America reveals that the pattern of "dominant opinion" did not favor an absolute free market. Holmes concluded that the dominant opinion was for an interpretation that upheld the New York state statute limiting working hours. It was only to a judge who had *laissez faire* economic beliefs that the statute would appear "novel, and even shocking." It was not a syllogism but the majority judges' economic beliefs—and accompanying sense of shock—that led them to strike the statute down.[6]

So a judge who expresses reasons for a judgment might not in truth be explaining his judgment. Such behavior is plainly at odds with the formal requirements of adjudication: the judge is supposed to say how he reaches his decisions. Holmes assumed that judges do not always do that. Their decisions are outputs derived from patterns found in experience,[7] not answers arrived at through logical proof. In Holmes's view, even legal texts, like constitutions, statutes, and past judgments, do not speak for themselves. As for artefacts of non-textual "experience"—the sources whose "significance is vital, not formal"[8]—those display their patterns even less obviously. Holmes thought that all the elements of experience, taken in aggregate, were the material from which derives the "judgment or intuition more subtle than any *articulate major premise*." It might not even "need research to show" what the premise is. How a decision-maker got from experience to decision—how the decision-maker found a pattern in the data, indeed even what data the decision-maker found the pattern in—thus remains unstated and thus obscure.

6.2   Machine Learning's
Inarticulate Major Premise

It is said—and it is the premise behind such regulatory measures as the GDPR—that machine learning outputs require explanation. Holmes's idea of the inarticulate major premise speaks directly to the problem of how to satisfy this requirement. Holmes said that the logic presented in a judicial decision to justify that decision was not an adequate explanation, and that for a full explanation one must look also to the body of experience that judges carry with them. For Holmes, the formal principle stated in a statute, and even in a constitutional provision, is not an adequate guide to the law, because to discern its proper meaning one must look at the traditions and opinions behind it.

Likewise, when considering the output of a machine learning system, the logic of its algorithms cannot supply an adequate explanation. We must look to the machine's "experience," i.e. to its training dataset.

One reads in Articles 13, 14, and 15 of the GDPR, the central loci of explainability, that meaningful information about an automated decision will come from disclosing "the *logic* involved."[9] This is a category error. A machine learning output cannot be meaningfully assessed as if it were merely a formula or a sum. A policymaker or regulator who thinks that machine learning is like that is like the unnamed judge whom Holmes made fun of for thinking a fault in a court judgment could be identified the way a mistake might be in arithmetic, or the named judges whom he said erred when they deduced that working hour limits on bakers are unconstitutional. The logic of deduction, in Holmes's idea of the law, is not where law comes from; it is certainly not in machine learning where outputs come from. The real source—the inarticulate major premise of law and of machine learning alike—is the data or experience.

If you follow Holmes and wish to explain how a law or judgment came to be, you need to know the experience behind it. If you wish to explain how a machine learning process generated a given output, you need to know the data that was used to train the machine. If you wish to make machine learning systems accountable, look to their training data not their code. If there is something that one does not like in the experience or in the data, then chances are that there will be something that one does not like in the legal decision or the output.

6.3 THE TWO CULTURES: SCIENTIFIC EXPLANATION VERSUS MACHINE LEARNING PREDICTION

To explain a decision, then, one must explain in terms of the data or experience behind the decision. But what constitutes a satisfactory explanation? In *Law in Science and Science in Law*, Holmes in 1899 opened the inquiry like this:

> What do we mean when we talk about explaining a thing? A hundred years ago men explained any part of the universe by showing its fitness for certain ends, and demonstrating what they conceived to be its final cause according to a providential scheme. In our less theological and more scientific day, we explain an object by tracing the order and process of its growth and development from a starting point assumed as given.[10]

Even where the "object" to be explained is a written constitution, Holmes said an explanation is arrived at "by tracing the order and process of its growth and development," as if the lawyer were a scientist examining embryo development under a microscope.[11] And yet for all of Holmes's scientific leaning, his best known epigram is expressed with an emphatically non-scientific word: "The *prophecies* of what the courts will do in fact, and nothing more pretentious, are what I mean by the law."[12]

Holmes seems to anticipate a tension that the philosophy of science has touched on since the 1960s and that the nascent discipline of machine learning since the 2000s has brought to the fore: the tension between explaining and predicting.

In the philosophy of science, as advanced in particular by Hempel,[13] an explanation consists of (i) an *explanans* consisting of one or more "laws of nature" combined with information about the initial conditions, (ii) an *explanandum* which is the outcome, and (iii) a deductive argument to go from the *explanans* to the *explanandum*. In fact, as described by Shmueli in his thoughtful examination of the practice of statistical modelling *To explain or to predict?*[14] it is actually the other way round: the goal of statistical modelling in science is to make inferences about the "laws of nature" given observations of outcomes. Terms like "law" and "rule" are used here. Such terms might suggest stipulation, like legal statutes, but in this context they simply mean scientific or engineering laws: they could be causal models[15] that aim to approximate nature, or they could simply be equations that describe correlations.

In the machine learning/prediction culture championed by Leo Breiman in his 2001 rallying call *Statistical modelling: the two cultures*, from which we quoted at the opening of Chapter 1, the epistemological stance is that explanation in terms of laws is irrelevant; all that matters is the ability to make good predictions. The denizens of the prediction culture sometimes have an air of condescension, hinting that scientists who insist on an explanation for every phenomenon are simpletons who, if they do not understand how a system works, can't imagine that it has any value. In the paper describing their success at ImageNet Challenge in 2012—following which the current boom in machine learning began—Krizhevksy et al. noted the challenge of getting past the gatekeepers of scientific-explanatory culture: "[A] paper by Yann LeCun and his collaborators was rejected by the leading computer vision conference on the grounds that it used neural networks and therefore provided no insight into how to design a vision system."[16] LeCun went on to win the 2018 Turing Award (the "Nobel prize for computer science") for his work on neural networks[17] and to serve as Chief AI Scientist for Facebook.[18]

Here is an illustration of the difference between the two cultures, as applied to legal outcomes. Suppose our goal is to find a formula for the probability that defendants will flee if released on bail: here we are inferring a rule, a formula that relates the features of objects under consideration to the outcomes, and which can be applied to any defendant. Or suppose our goal is to determine whether the probability is higher for violent crime or for drug crime all else being equal: here again we are making an inference about rules (although this is a more subtle type of inference, a comparative statement about two rules which does not actually require those rules to be stated explicitly).

By contrast, suppose our goal is to build an app that estimates the probability that a given defendant will flee: here we are engaging in prediction.[19] We might make a prediction using syllogistic inference, or by reading entrails, or with the help of machine learning. The distinguishing characteristic of prediction is that we are making a claim about how some particular case is going to go.

Making a prediction about a particular case and formulating a rule of general application are tightly interwoven. Their interweaving is visible in judicial settings. One of Holmes's Supreme Court judgments is an example. Typhoid fever had broken out in St. Louis, Missouri. The State of Missouri sued Illinois, on the theory that the outbreak was caused by a recent change in how the State of Illinois was managing the river at the city of Chicago. In *State of Missouri* v. *State of Illinois* Holmes summarized Missouri's argument as follows:

> The plaintiff's case depends upon an inference of the unseen. It draws the inference from two propositions. First, that typhoid fever has increased considerably since the change, and that other explanations have been disproved; and second, that the bacillus of typhoid can, and does survive the journey and reach the intake of St. Louis in the Mississippi.[20]

In support of this second proposition, Missouri put forward rules, formulated with reference to its experts' observations, stating how long the typhoid bacillus survives in a river and how fast the Mississippi River might carry it from Chicago to St. Louis. If you accept the rules that Missouri formulated from its experts' observations, then you could express the situation like this:

> Let x = miles of river between downstream location of outbreak and upstream location of a typhoid bacillus source.

> Let y = rate in miles per day at which typhoid bacillus travels downstream in the river.

> Let z = maximum days typhoid bacillus survives in the river.

> If $x \div y \leq z$, then the bacillus survives—and downstream plaintiff wins;

> If $x \div y > z$, then the bacillus does not survive—and downstream plaintiff loses.

Expressing the situation this way necessarily has implications for other cases. Justice Holmes drew attention to the implications: the winning formula for Missouri as plaintiff against Illinois might well later have been a losing one for Missouri as defendant against a different state. "The plaintiff," Holmes wrote, "obviously must be cautious upon this point, for if this suit should succeed, many others would follow, and it not improbably would find itself a defendant to a [suit] by one or more of the states lower down upon the Mississippi."[21]

Missouri was making an inference of the unseen *in a particular instance*, which in machine learning terminology is referred to as prediction. Missouri used general propositions to support this prediction, and Holmes (with his well-known distrust of the general proposition) warned that such reasoning can come back to bite the plaintiff.

The difference between finding rules and making predictions might seem slight. If we have a rule, we can use it to make predictions about future cases; if we have a mechanism for making predictions, that mechanism may be seen as the embodiment of a rule. Hempel did not see any great difference between explanation and prediction. To Hempel, an explanation is after the fact, a prediction is before the fact, and the same sort of deductive reasoning from natural laws applies in both cases.

But what if it is beyond the grasp of a simple-minded philosopher— or, for that matter, of any human being—to reason about the predictive mechanism? This is the real dividing line between the two cultures. The scientific culture is interested in making inferences about rules, hence *a fortiori* practitioners in the scientific culture will only consider rules of a form that can be reasoned about. The prediction culture, by contrast, cares about prediction accuracy, even if the prediction mechanism is so complex it seems like magic.

Arthur C. Clarke memorably said, "Any sufficiently advanced technology is indistinguishable from magic."[22] Clarke seems to have been thinking about artefacts of a civilization more advanced than that of the observer trying to comprehend them. Thus, a stone age observer, presented with a video image on a mobile phone, might think it magical. It would take more than moving pictures to enchant present-day observers, but we as a society have built technological artefacts whose functioning we struggle to explain.

The prediction culture says that we should evaluate an artefact, even one that seems like magic, by whether or not it actually works. We can still make use of machines that embody impenetrable mechanisms; we should evaluate them based on black-box observations of their predictive accuracy. A nice illustration may be taken from a case from the U.S. Court of Appeals for the 7th Circuit in 2008. A company had been touting metal bracelets. The company's assertions that the bracelets were effective as a cure for various ailments were challenged as fraudulent. Chief Judge Easterbrook, writing for the 7th Circuit, recalling the words of Arthur C. Clarke that we've just quoted above, was dubious about "a person who promotes a product that contemporary technology does not understand"; he said that such a person "must establish that this 'magic' actually works. Proof is what separates an effect new to science from a swindle."[23] Implicit here is that the "proof," while it might establish *that* the "magic" works, does not necessarily say anything about *how* it works. Predicting and explaining are different operations. Easterbrook indeed goes on to say that a placebo-controlled, double-blind study—that is, the sort of study prescribed by the FDA for testing products that somebody hopes to market as having medical efficacy—is "the best test" as regards assertions of medical efficacy of a product.[24] Such a test, in itself, solely measures outputs of the (alleged) medical device; it is not "proof" in the sense of a mathematical derivation. It does not require any understanding of how the mechanism works; it is just a demonstration that it *does* work. True, a full-scale FDA approval process—a process of proof that is centered around the placebo-controlled, double-blind study that the judge mentions—also requires theorizing as to how the mechanism works, not just black-box analysis. But Easterbrook here, focusing on proof of efficacy, makes a point much along the lines of Breiman: a mechanism can be evaluated purely on whether one is satisfied with its outcomes, rather than on considerations such as parsimony or interpretability or consonance with theory.[25] A mechanism can be evaluated by seeking to establish whether "this 'magic' actually works."

Holmes made clear his view that a judicial explanation is really an apologia rather than an explanation, and that the real explanation is to be found by looking for the "inarticulate major premise" that comes from the jurist's body of experience. Holmes shied away from asking for logical or scientific explanations as a way to understand the jurist's experience. He instead invoked *prophecy*. Holmes went beyond logic (because simple mathematical arguments are inadequate), and beyond scientific explanation (perhaps because such explanation would either be inaccurate or incomprehensible when applied to jurists' behavior), and he came finally to prediction. In this, Holmes anticipated machine learning.

6.4   Why We Still Want Explanations

The inarticulate major premise, starting immediately after Holmes's dissent in *Lochner*, provoked concern, and it continues to.[26] Unexplained decisions, or decisions where the true reasons are obscured, are inscrutable, and, therefore, the observer has no way to tell whether the reasons are valid. Validity, for this purpose, may mean technical correctness; it also may mean consonance with basic values of society. Testing validity in both these senses is an objective behind explainability. We turn here in particular to values.[27]

Eminent readers of Holmes conclude that he didn't have much to say about values.[28] But he was abundantly clear that, whatever the values in society might be, if they form a strong enough pattern, then they are likely to find expression in law: "Every opinion tends to become law."[29] Whether or not one has an opinion about the opinion that becomes law, Holmes described a process that has considerable present-day resonance. Data from society at large will embody opinions held in society at large; and thus a machine learning output derived from a pattern found in the data will itself bear the mark of those opinions.

The influence of opinions held in society would be quite straightforward if there were no conflicting opinions. But many opinions do conflict. Holmes plainly was concerned about discordance over values; it was to accommodate "fundamentally differing views" that he said societies adopt constitutions.[30] Less clear is whether he thought that certain values are immutable, imprescriptible, or in some fashion immune to derogation. He suggested that some might be: he said that a statute might "infringe fundamental principles." He didn't say what principles might be fundamental.

A law, if it embodied certain biases held in society, would infringe principles held to be fundamental today. Examples include racial and gender bias. In Holmes's terms, those are "opinions" that should not "become law." Preventing them from becoming law is a central concern today. The concern arises, *mutatis mutandis*, with machine learning outputs. Where machine learning outputs have legal effects, they too will infringe fundamental principles, if they embody biases such as racial or gender bias. Preventing such "opinions" from having such wider influence is one of the main reasons that policy makers and writers have called for explainability.

In short, in both processes, law and machine learning, the risk exists that experience or data shaped a decision that ought not have been allowed to.[31] In both, however, the experience or the data might not be readily visible.[32] As we will explore in Chapters 7 and 8, much of the concern over its potential impact on societal values relates to this obscurity in machine learning's operation.

NOTES

1. For literature see, e.g., Casey, *The Next Chapter in the GDPR's "Right to Explanation" Debate and What It Means for Algorithms in Enterprise*, EUROPEAN UNION LAW WORKING PAPERS, No. 29 (2018) and works cited id., at p. 14 n. 41.
2. See Grant & Wischik, *Show Us the Data: Privacy, "Explainability," and Why the Law Can't Have Both*, forthcoming, 88 GEO. WASH. L. REV. (Nov. 2020). See also, positing a conflict between privacy and data protection regulations, on the one hand, and anti-discrimination regulations on the other, Žliobaitė & Custers (2016).
3. In Aristotle's logic, the "major premise" is a stated element at the starting point of a syllogism. See Robin Smith, *Aristotle's Logic*, in Zalta (ed.), THE STANFORD ENCYCLOPEDIA OF PHILOSOPHY (Summer 2019 edn.): https://plato.stanford.edu/entries/aristotle-logic/.
4. Holmes, *The Theory of Legal Interpretation*, 12 HARV. L. REV. 417, 420 (1898–1899).
5. *Lochner v. New York*, 198 U.S. 45, 75–76, 25 S.Ct. 539, 547 (Holmes, J., dissenting, 1905).
6. Differences of interpretation exist among jurists reading the passage in Holmes's *Lochner* dissent about "[g]eneral propositions" and judgments or intuitions "more subtle than any articulate major premise." No less an authority on Holmes than Judge Posner once referred to the passage to mean that certain "statements should be treated as generalities open to exception": *Arroyo v. U.S.*, 656 F.3d 663, 675 (Posner, J., concurring, 7th Cir., 2011). We read it to mean something more. It means that reasons that in truth led to a judicial outcome are sometimes not expressed in the judgment. That is, stated reasons in a judgment, which typically take the form of a logical proof proceeding to the judge's conclusion from some major premise that the judge has articulated, are not the real explanation for why the judge concluded the way he did. Our reading accords with a train of thought running through Holmes's work, at least as far back as THE COMMON LAW (1881). A number of judges have read the passage as we do: See, e.g., *City of Council Bluffs v. Cain*, 342 N.W.2d 810, 814 (Harris, J., Supreme Court of Iowa, 1983); *Loui v. Oakley*, 438 P.2d 393, 396 (Levinson, J., Supreme Court of Hawai'i, 1968); *State v. Farrell*, 26 S.E.2d 322, 328 (Seawell, J., dissenting, Supreme Court of North Carolina, 1943).
7. Such experience not infrequently includes implicit bias. For an example, see Daniel L. Chen, Yosh Halberstam, Manoj Kumar & Alan C. L. Yu, *Attorney Voice and the US Supreme Court*, in Livermore & Rockmore (eds.) (2019) p. 367 *ff.*
8. Gompers v. United States, 233 U.S. 604, 610 (1914).

9. Emphasis ours. See also in the GDPR Arts. 21–21 and Recital 71. One reads in the legal scholarship, too, that it is because some algorithms are "more complex" than others that they are harder to explain. See, e.g., Hertza, 93 N.Y.U. L. REV. 1707, 1711 (2018). Mathematical complexity of algorithms is not what drives machine learning, however. Data is. See Chapter 3, pp. 35–38.

10. *Law in Science and Science in Law*, 12 HARV. L. REV. at 443 (1898–1899).

11. Cf. Holmes's description of a constitution as "the skin of a living thought", Chapter 4, p. 47, n. 7.

12. *Path of the Law*, 10 HARV. L. REV. at 461 (1896–1897).

13. See, for example, *Scientific Explanation*, from the STANFORD ENCYCL. PHILOS. (Sept. 24, 2014): https://plato.stanford.edu/entries/scientific-explanation/. Of relevance here is the inductive-statistical model, due to Hempel (1965).

14. Galit Shmueli, *To Explain or to Predict?* 25(3) STAT. SCI. 289–310 (2010).

15. In the social sciences, inference, especially inference about causal relationships, is typically preferred to prediction. But for a defense of prediction see Allen Riddell, *Prediction Before Inference*, in Livermore & Rockmore (ed.) (2019) 73–89. See also Breiman, *The Two Cultures*, quoted above, Chapter 1, p. 1.

16. Krizhevksy, Sutskever & Hinton (2017).

17. https://amturing.acm.org/award_winners/lecun_6017366.cfm.

18. https://www.linkedin.com/in/yann-lecun-0b999/. Retrieved 19 April 2020.

19. Kleinberg et al. note that judges are supposed by law to base their bail decision solely on this prediction, and they show that a machine learning algorithm does a better job. Kleinberg, Lakkaraju, Leskovec, Ludwig & Mullainathan, *Human Decisions and Machine Predictions*, 46 Q. J. ECON. 604–32 (2018).

20. *State of Missouri v. State of Illinois*, 26 S.Ct. 270, 200 U.S. 496, 522–23 (1906).

21. 200 U.S. at 523.

22. Clarke (1962) 21.

23. FTC v. QT, Inc., 512 F.3d 858, 862 (7th Cir. 2008).

24. Id.

25. See Chapter 1, pp. 10–11.

26. Since 1917 when Albert M. Kales, *"Due Process," the Inarticulate Major Premise and the Adamson Act*, 26 YALE L. J. 519 (1917), addressed Holmes's famous *Lochner* dissent, over a hundred American law review articles have addressed the same. It has concerned lawyers in Britain and the Commonwealth as well: see, e.g., the Editorial Notes of the first issue

of MODERN LAW REVIEW: 1(1) MLR 1, 2 (1937). Writings on the matter are recursive: see Sunstein, *Lochner's Legacy*, 87 COL. L. REV. 873–919 (1987); Bernstein, *Lochner's Legacy's Legacy*, 82 TEX. L. REV. 1–64 (2003).

27. For some recent work on the challenge of getting AI to reflect social values in legal operations, see Al-Abdulkarim, Atkinson & Bench-Capon, *Factors, Issues and Values: Revisiting Reasoning with Cases*, International Conference on AI and Law 2015, June 8–12, 2015, San Diego, CA: https://cgi.csc.liv.ac.uk/~tbc/publications/FinalVersionpaper44.pdf.

28. Most comprehensively, see Alschuler (2000). See also, e.g., Jackson, 130 HARV. L. REV. 2348, 2368–70 (2017).

29. *Lochner* (Holmes, J., dissenting), op. cit.

30. Id.

31. Kroll et al. (op. cit.) described the matter in regard to machine learning like this:

> machine learning can lead to discriminatory results if the algorithms [sic] are trained on historical examples that reflect past prejudice or implicit bias, or on data that offer a statistically distorted picture of groups comprising the overall population. Tainted training data would be a problem, for example, if a program to select among job applicants is trained on the previous hiring decisions made by humans, and those prevision decisions were themselves biased. 165 U. PA. L. REV. at 680 (2017).

Barocas & Selbst, 104 CAL. L. REV. at 674 (2016), to similar effect, say, "[D]ata mining can reproduce existing patterns of discrimination, inherit the prejudice of prior decision makers, or simply reflect the widespread biases that persist in society." Cf. Chouldechova & Roth, *The Frontiers of Fairness in Machine Learning*, Section 3.3, p. 6 (Oct. 20, 2018): https://arxiv.org/pdf/1810.08810.pdf.

32. See generally Pasquale (2015). Though the emphasis in the 2015 title on algorithms is misplaced, Pasquale elsewhere has addressed distinct problems arising from machine learning: Pasquale (2016).

CHAPTER 7

Juries and Other Reliable Predictors

In the formalist's account, a jury settles factual questions; only the judge settles questions of law.[1] Holmes thought that juries exercise a wider influence than that.[2] "I don't like to be told that I am usurping the functions of the jury if I venture to settle the standard of conduct myself in a plain case," Holmes wrote in dryly humorous vein to his long-time friend Frederick Pollock. "Of course, I admit that any really difficult question of law is for the jury, but I also don't like to hear it called a question of fact..."[3] Whatever the proper way to describe the division of responsibilities between judge and jury, Holmes saw the latter to be a sort of conduit into the courtroom of the general understandings—and feelings—prevalent in the community. In its best aspect, the jury brought the collective experience of the community to questions that best are settled through the application of common sense. The jury, as an embodiment of community experience, is called upon when what matters is "the nature of the act, and the kind and degree of harm done, considered in the light of expediency and usage."[4] However, the jury is involved in several vexing problems. Analogous problems arise with machine learning.

7.1 Problems with Juries, Problems with Machines

Juries, wrote Holmes, "will introduce into their verdict a certain amount—a very large amount, so far as I have observed—of popular prejudice."[5] True, juries "thus keep the administration of law in accord with

© The Author(s) 2020
T. D. Grant and D. J. Wischik, *On the path to AI*,
https://doi.org/10.1007/978-3-030-43582-0_7

the wishes and feelings of the community."[6] Reading this "accord with" clause on its own, one might think it flattery. Reading in context, one sees that Holmes had no intention to flatter juries. Addressing why we have juries at all, Holmes added,

> [S]uch a justification is a little like that which an eminent English barrister gave me many years ago for the distinction between barristers and solicitors. It was in substance that if law was to be practised somebody had to be damned, and he preferred that it should be somebody else.[7]

Holmes, in stating that juries produce results "in accord with the wishes and feelings of the community," was stating a fact about the behavior of juries as he had observed many times juries to behave. He was not conferring a blessing upon them for behaving that way.

Three particular problems with juries suggested to Holmes a need for caution. In each a similar problem may be discerned with machine learning.

First, there is a problem of accountability. Holmes, in referring to the barrister who preferred that "somebody else" bear the blame, was referring to the practice of English advocates not to make a point about advocacy but to make a point about decision-making. If a judge is faced with an intractable question about which he would really prefer not to make a decision but must for purposes of deciding the case, he looks for a way to send that question to the jury. Some decisions are unlikely to please all parties concerned. Some decisions are likely, instead, to provoke criticism and resistance. If an authority, such as a judge, has the discretion to devolve such decisions upon another actor, then the temptation exists to do so, because if the other actor makes the decision then the authority removes himself from blame.

A temptation to devolve decisions exists with machine learning. The role of machine learning systems in actual decision practice grows apace. The growth is visible, or foreseeable, in banks,[8] on highways,[9] on the battlefield,[10] in the courtroom.[11] Instead of a human being making the decision—such as a bank loan officer or a sentencing judge—the decision is given to a machine. A certain distance now separates the human being from the decision and its consequences. If "somebody had to be damned," e.g. for denying home mortgages or giving long jail sentences on invidious criteria, then authorities and their institutions might well like to see that distance increase. Given enough distance, the machine makes the decision and is, perhaps, to be damned; but the human being holds up his hands and declares, *Don't look at me.*

So far, decision making by machine has not relieved human beings of the duty, where the law imposes it, to give account for actions that they set in train. It has however started to raise questions of causation.[12] Those questions are likely to multiply if the separation continues to increase (as it probably will) between a human being setting a machine process in train and the practical impact of the machine decision. Proposals, such as that in the European Parliament in 2017 to confer legal personality on machines,[13] would pave the way to even further separation (and for that reason, among others, are a bad idea).[14] We find a timely message in Holmes's caution toward the devolution of decisions to juries.

Second, there is propagation of "popular prejudice." Just as the jury is faithful to the experience it brings to the court room and thus delivers a verdict that reflects the patterns in that experience, so will the machine be faithful to its training data. If the training data embody a pattern of community prejudice that we do not wish to follow, then we will not be pleased with the decision that the machine delivers.[15] Holmes viewed juries as a mechanism to transmit community experience into legal decision. Affirming that they are reliable in that function is to concede a certain admiration, but it is just as much to sound a warning. No less reliable is the machine learning system in producing an output using the data that it is given. Both are mechanisms that rely on their givens—their experience and the data. Their outputs necessarily reflect the patterns they find therein.[16]

Finally, there is the problem of how to scrutinize black-box decision making. When policy-makers call for machine decision making to be scrutinized, they encounter a difficulty resembling that which arises when we try to figure out how a jury reached its decision: it is difficult to see inside the decision making mechanism. In Chapter 2,[17] we addressed the challenges involved in explaining the outputs of black-box systems. Holmes noted the challenges in testing juries and similar bodies. In considering a decision by a state taxation tribunal, which because that tribunal was constituted of laypersons was like a jury in the relevant way, Holmes wrote, "how uncertain are the elements of the evidence, and in what unusual paths it moves"[18]; an appellate court must "mak[e] allowance for a certain vagueness of ideas to be expected in the lay mind."[19] Even the judge, thought Holmes, tended, "[w]here there is doubt..." to produce decisions derived from reasons that are "disguised and unconscious."[20] Even the judge, schooled in the formalities of logic, produces outputs that are inscrutable. The jury is all the less likely to supply a clear trace of what led it to decide as it did. Here again is the phenomenon of occluded reasoning, the hidden layers of legal process. Explaining a machine learning output presents in a new setting a problem familiar since at least Holmes's day in law.

7.2   WHAT TO DO ABOUT THE PREDICTORS?

Holmes, after he came to be cast in the role of progenitor of legal realism, was widely described as complacent about social problems, callous, even, in the face of injustice. Present-day observers have gone so far as to call Holmes "corrosive" because he seemed to accept the state of affairs as it was.[21] In one judgment, which his critics have often cited, Holmes upheld a statute of Virginia under which the Commonwealth sterilized certain persons deemed "feebleminded."[22] Holmes supplied personal material, too, that later observers would use to characterize him as cold or resigned. Writing in 1927 in a letter to his friend Harold Laski, Holmes said "I do accept 'a rough equation' between isness and oughtness."[23] To Morris Cohen, he wrote in 1921, "I do in a sense worship the inevitable."[24] Writing around the same time to another correspondent, Holmes doubted that rational improvements would be made in the world at least in the immediate future:

> We all try to make the kind of a world that we should like. What we like lies too deep for argument and can be changed only gradually, often through the experience of many generations.[25]

One might read Holmes to counsel acceptance of whatever prediction the patterns in past experience suggest. And, yet, Holmes took account of the desire for change—the effort by all to "make the kind of a world that we should like." He also showed personal interest in members of the younger generation who sought change; the correspondent to whom he wrote about the kind of a world that we should like was John C. H. Wu, a 22 year old law student at the time and much concerned with lifting China, his native land, out of its then century-long malaise.[26] The standard account—of Holmes as fatalist—is not supported by his record of encouraging people such as Wu in their efforts to escape the limits of experience.[27]

Moreover, concerning experience and its influence on decision, Holmes did not think that all mechanisms for discerning patterns in experience merit equal deference. The legislature was the mechanism, in Holmes's understanding, that a court was least to question; deference to the legislature was a precept that Holmes faithfully applied. It was that deference that is visible in his judgment in the Virginia sterilization case. One sees it as well in his dissents in *Lochner* and similar cases, where he decried the Court for second-guessing laws that had been enacted under proper legislative procedures and that aimed at various purposes that today would be called socially progressive.

In a number of appeals from jury verdicts, Holmes had a very different response than he had to challenges against statutes. He overturned jury verdicts or dissented against majorities that didn't. *Moore v. Dempsey* is the most prominent of the cases in which Holmes considered how a jury, perfectly "in accord with the wishes and feelings of the community," produced a result that demanded correction. Five defendants, all African American, had been arrested. The grand jury that returned indictments against them had been comprised of whites only, including the members of an attempted lynch mob. The evidence against them was scarce; their defense lawyers, appointed by the court, had done nothing in their defense; and the "Court and the neighborhood were thronged with an adverse crowd that threatened the most dangerous consequences to anyone interfering with the desired result."[28] The jury at trial found the defendants guilty; the sentence was death. In Holmes's words, "counsel, jury and judge were swept to the fatal end by an irresistible wave of public passion."[29] True, the situation in *Moore v. Dempsey* had been that the facts "if true as alleged... [made] the trial absolutely void."[30] The body assembled to function as a jury had not functioned as a jury at all. At the same time, the transactions in the courthouse exemplified, *in extremis*, the jury as decision-maker. Extreme example though it was, the case shone a light on how the jury works, and juries work that way even when all persons involved have acted in good faith: juries find their patterns in the experience around them. A corrective is called for, when that experience discords with our better understanding of the world we wish to have.

As the data that trains the machine learning system is a given, both in Latin grammar and in the process of machine learning, so too is the experience that Holmes understood to be the main influence on law. Holmes's personal outlook was congenial to accepting givens. However, Holmes was alert to the danger that givens present for certain kinds of decision-making that rely upon them as their inputs. He didn't suggest that the jury could do any differently. He placed the corrective someplace else, namely in the hands of the court of appeal. We will turn in the next chapter to consider more closely how Holmes understood the corrective to work in a particular situation—that where public authorities had garnered evidence in breach of the constitutional right of a defendant—and how a practice we will call *inferential restraint* has been necessary in legal procedures and is likely to be in machine learning as well.

NOTES

1. For a classic account, see Leonard (ed.), THE NEW WIGMORE (2010) § 1.2 pp. 3–5.
2. Holmes had experience with juries. He judged many appeals from trials in which they were empaneled. Moreover, for most of the time when Holmes served on the Massachusetts Supreme Judicial Court, its members also sat as trial judges (in divorces, murders, certain contractual disputes, contests over wills and trusts, suits in equity). Budiansky 183–84. Further to Holmes's jury trials, see Zobel, 8 MASSACHUSETTS LEGAL HISTORY 35–47 (2002); 36 BOSTON BAR J. 25–28 (1992).
3. *Letter of Holmes to Pollock* (May 13, 1898): reprinted De Wolfe Howe (ed.) (1942) 85–86 (ellipses original). Twenty-first century writers also have doubted the classic distinction: see, e.g., Allen & Pardo, *The Myth of the Law-Fact Distinction*, 97 Nw. U. L. REV. 1769–1807 (2003). See also Zuckerman, *Law, Fact or Justice?* 66 BOSTON U. L. REV. 487 (1986).
4. Middlesex v. McCue, 149 Mass. 103; 21 N.E. 230 (1889). Cf. Commonwealth v. Perry, 139 Mass. 198, 29 N.E. 356 (1885).
5. *Law in Science and Science in Law*, 12 HARV. L. REV. at 460.
6. Id.
7. Id. In England and Wales and in other countries of the Commonwealth, the practice of law is divided between solicitors and barristers. Broadly speaking, the division is between lawyers who don't argue in court (solicitors) and those who do (barristers). The exclusivity of the barristers' rights of appearance in court that prevailed in Holmes's day has been qualified since, but the basic division remains the same. For an account of the relationship by a UK law firm in 2016, see https://www.slatergordon.co.uk/media-centre/blog/2016/09/difference-between-a-lawyer-a-solicitor-and-a-barrister-explained/.
8. Bartlett, Morse, Stanton & Wallace, *Consumer-Lending Discrimination in the Era of FinTech* (Oct. 2018). http://faculty.haas.berkeley.edu/

morse/research/papers/discrim.pdf. Cf., considering a different problem of accountability, Ji, *Are Robots Good Fiduciaries: Regulating Robo-Advisors Under the Investment Advisers Act of 1940*, 117 COL. L. REV. 1543 (2017).

9. As to a possible tort law solution, see Geistfeld, *A Roadmap for Autonomous Vehicles: State Tort Liability, Automobile Insurance, and Federal Safety Regulation*, 105 CAL. L. REV. 1611 (2017). Cf. Kowert, *Note: The Foreseeability of Human-Artificial Intelligence Interactions*, 96 TEX. L. REV. 181 (2017).

10. See Ford, *AI, Human-Machine Interaction, and Autonomous Weapons: Thinking Carefully About Taking 'Killer Robots' Seriously*, Arms Control and International Security Papers, 1(2) (April 20, 2020). Cf. Feickert et al., *U.S. Ground Forces Robotics and Autonomous Systems (RAS) and Artificial Intelligence (AI)* (Congressional Research Service Reports, Nov. 1, 2018).

11. See for example Roth, *Trial by Machine*, 104 GEO. L. J. 1243 (2016). Cf. Rolnick Borchetta, *Curbing Collateral Punishment in the Big Data Age: How Lawyers and Advocates Can Use Criminal Record Sealing Statutes to Protect Privacy and the Presumption of Innocence*, 98 B.U. L. REV. 915 (2018).

12. As to causation and challenges in dealing with legal liability for machine outputs, see Chapter 8, p. 97 n. 12.

13. Civil Law Rules on Robotics, European Parliament Resolution (Feb. 16, 2017).

14. For objections to proposals to confer legal personality on machine learning systems and the like, see Bryson, Diamantis & Grant (2017). See also Brożek & Jakubiec (2017).

15. A point made by Kroll et al., *supra* n. 76 at 687.

16. Considering this similarity, one might be cautious about using a machine learning system to "debias" a jury. We are not talking here about using machines such as polygraphs as supplements to decision-making by a jury; we express no view here about such machines. As to "debiasing juries" with machines, see Roth op. cit. at 1292–94.

17. See Chapter 2, pp. 24–27.

18. Coulter et al. v. Louisville & Nashville Railroad Company, 196 U.S. 599, 609 25 S.Ct. 342, 344 (Holmes, J.) (1905).

19. 196 U.S. at 610, 25 S.Ct. at 345.

20. 12 HARV. L. REV. at 461.

21. Jackson, op. cit. 130 HARV. L. REV. at 2368 (2017).

22. *Buck v. Bell*, 274 U.S. 200 (Holmes, J., 1927). As to the circumstances of the case and the drafting of Holmes's judgment see Budiansky 428–430.

23. Letter from Holmes to Laski (June 1, 1927), reprinted in 2 HOLMES-LASKI LETTERS 948 (Mark De Wolfe Howe ed., 1953).

24. Letter from Holmes to Morris Cohen (Jan. 30, 1921), in Felix S. Cohen, *The Holmes-Cohen Correspondence* (1948) 9 J. HIST. IDEAS 3, 27.
25. Letter from Holmes to Wu, December 12, 1921, reprinted in JUSTICE HOLMES TO DOCTOR WU: AN INTIMATE CORRESPONDENCE, 1921–1932 (1947) 2–3.
26. Wu, who was studying at the University of Michigan and approached the 80-year-old Associate Justice evidently on self-introduction, maintained a long correspondence with Holmes. Wu went on to have a distinguished career as academic, lawyer, and diplomat.
27. There were others noted for their activism and whom Holmes nevertheless admired and had long friendships with. Felix Frankfurter and Louis Brandeis (who both in time served as Associate Justices of the U.S. Supreme Court, Brandeis overlapping with Holmes's tenure, Frankfurter taking a seat later) were prominent examples. See Budiansky 328–29, 387–90; 356–59, 362–63.
28. Moore v. Dempsey, 261 U.S. 86, 89 (Holmes, J., 1923).
29. 261 U.S. at 91.
30. 261 U.S. at 92.

CHAPTER 8

Poisonous Datasets, Poisonous Trees

As we addressed in preceding chapters,[1] data, or "experience" as Holmes referred to the inputs in law, influences decision-making in a number of ways. It might influence decision-making in such a way that the decision made is illegal, immoral, unethical, or undesirable on some other grounds. Both legal decision-making and machine learning have struggled about what to do when presented with data that might influence decision-making in such a way. Courts have attempted to exclude particular pieces of data, what we will call "bad evidence," from the decision-making process altogether.[2] Exclusion before entry into the process removes the problem: data that doesn't enter doesn't affect the process. However, exclusion also may *introduce* a problem. In both legal settings and in machine learning, particular pieces of evidence or data might have undesired effects, but the same inputs might assist the process or even be necessary to it. It also might be that no mechanism exists that will reliably exclude only what we aim to exclude. So exclusion, or its collateral effects, may erode the efficacy or integrity of the process. Lawyers refer to the "probative value" of a piece of evidence, an expression they use to indicate its utility to the decision process—even when a risk exists that that evidence might have undesired effects.[3] The vocabulary of machine learning does not have such a received term here, but data scientists, as we described in Chapter 3, know well that the data that trains the machine is essential to its operation.

Exclusion is not the only strategy courts have used to address bad evidence and its kindred problem, bias. Another is to restrain the inferences

© The Author(s) 2020
T. D. Grant and D. J. Wischik, *On the path to AI*,
https://doi.org/10.1007/978-3-030-43582-0_8

that the decision-maker draws from certain evidence that might otherwise have undesirable effects on decision-making. This strategy entails an adjustment to the inner workings of the process of decision itself.

Finally, restraint may be imposed at a later stage. For example, courts review outputs (verdicts and judgments) and, if they're not in accord with certain rules, strike them down, which, in turn, means that the instruments of public power will not act on them. In that strategy, a decision-making mechanism—for example, a jury, and as much might be said of a machine learning system—was not under inferential restraint (or it was but it ignored the restraint); the output the mechanism gives, on review, is unacceptable in some way; and, so, the output is not used. The restraint did not operate within the mental or computational machinery that generated an output but, instead, upon those persons or instrumentalities who otherwise would have applied the output in the world at large. Defect in the output discerned, they don't apply it.

We turn now to consider more closely the problem of bad evidence; the limits of evidentiary exclusion as a strategy for dealing with bad evidence in machine learning; and the possibility that restraining the inferences drawn from data and restraining how we use the outputs that a machine reaches from data—strategies of restraint that have antecedents in jurisprudence—might be more promising approaches to the problem of bias in the machine learning age.

8.1   THE PROBLEM OF BAD EVIDENCE

As an Associate Justice of the Supreme Court, Holmes had occasion in *Silverthorne Lumber Co. v. United States*[4] to consider a case of bad evidence. Law enforcement officers had raided a lumber company's premises "without a shadow of authority" to do so.[5] It was uncontested that, in carrying out the raid and taking books, papers, and documents from the premises, they had breached the Fourth Amendment, the provision of the United States Constitution that protects against unreasonable searches and seizures. The Government then sought a subpoena which would authorize its officers to seize the documents which they had earlier seized illegally. Holmes, writing for the Supreme Court, said that "the knowledge gained by the Government's own wrong cannot be used by it in the way proposed."[6] In result, the Government would not be allowed to use the documents[7]; it would not be allowed to "avail itself of the knowledge obtained by that means."[8] Obviously, no judge could efface the knowledge actually gained and thus lodged in the minds of the government officers concerned. The solution was instead to place a limit on what those officers were permitted to do with the knowledge: they were forbidden from using it to evade the original exclusion.

The "fruit of the poisonous tree," as the principle of evidence applied in *Silverthorne Lumber Co.* came to be known, is invoked in connection with a range of evidentiary problems. Its distinctiveness is in its application to "secondary" or "derivative" evidence[9]—i.e., evidence such as that obtained by the Government in *Silverthorne Lumber* on the basis of evidence that had earlier been excluded. *Silverthorne* and, later, *Nardone v. United States*, where Holmes's friend Felix Frankfurter gave the principle its well-known name, concerned a difficult question of causation. This is the question, a recurring one in criminal law, whether a concededly illegal search and seizure was really the basis of the knowledge that led to the acquisition of new evidence that the defendant now seeks to exclude. In the second *Nardone* case, Justice Frankfurter writing for the Court reasoned that the connection between the earlier illegal act and the new evidence "may have become so attenuated as to dissipate the taint."[10] But if the connection is close enough, if "a substantial portion of the case against him was a fruit of the poisonous tree,"[11] then the defendant, as of right, is not to be made to answer in court for that evidence.[12] That evidence, if linked closely enough to the original bad evidence, is bad itself.

We have noted three strategies for dealing with bad evidence: one of these is to cut out the bad evidence and so prevent it from entering the decision process in the first place. This strategy, which we will call *data pruning*,[13] in a judicial setting is to rule certain evidence inadmissible. It is a complete answer, when you have an illegal search and seizure, to the question of what to do with the evidence the police gained from *that* search. You don't let it in. A different strategy is called for, however, if bad evidence already has entered some phase of a decision process. Judges are usually concerned here with the jury's process of fact-finding. On close reading, one sees that Holmes in *Silverthorne Lumber* was concerned with the law enforcement officers' process of investigation. In regard to either process, and various others, a strategy is called for that restrains the inferences one draws from bad evidence. We will call that strategy *inferential restraint*. Finally, and further down the chain, where a decision or other output might be turned into practical action in the world at large, a further sort of restraint comes into play: restraint upon action. We will call this variant of restraint *executional restraint*. Data pruning and the two variants of restraint, all familiar since Holmes's day in American court rooms, have surfaced as possible strategies to address the problems that arise with data in machine learning. We will suggest, given the way machine learning works, that data pruning and strategies of restraint are not equally suited to address those problems.

8.2   DATA PRUNING

Excluding bad evidence from a decision process has at least two aims. For one, it has the aim of deterring impermissible practices by those who gather evidence, in particular officials with police powers. Courts exclude evidence "to compel respect for the constitutional guaranty [i.e., against warrantless search and seizure] in the only effectively available way—by removing the incentive to disregard it."[14] For another, it has the aim of preventing evidence from influencing a decision, if the evidence tends to produce unfair prejudice against the party subject to the decision. In machine learning, the first of these aims—deterring impermissible data-gathering practices—is not absent. It is present in regulations on data protection.[15] Our main focus here is with the second aim: preventing certain data from influencing the decision.[16] Data pruning is the main approach to achieving that aim in judicial settings.

Data pruning avoids thorny questions of logic, in particular the problem of attenuated causation. Just what inferences did the jury draw from the improper statements or evidence? Just what inferences did the police draw from the evidence gained from the unlawful search? And how did any such inferences affect future conduct (meaning future decision)? It is better not to have to ask those questions. This is a salient advantage of data pruning. It obviates asking, as Justice Frankfurter had to, whether the link between the bad evidence and the challenged evidence has "become so attenuated as to dissipate the taint."[17]

Data pruning has the related advantage that, if the bad data is cut away before the decision-maker learns of it, the decision-maker does not have to try not thinking about something that she already knows. Data pruning avoids the problem that knowledge gained cannot be unlearnt. As courts have observed, one "cannot unring a bell."[18] The cognitive problem involved here is also sometimes signaled with the command, "Try not to think of an elephant." By deftly handling evidentiary motions, or where needed by disciplining trial counsel,[19] the judge cuts out the elephant before anybody has a chance to ask the jury not to think about it.

Machine learning has a fundamental difficulty with data pruning. To make a meaningful difference on the learnt parameters, and thus on the eventual outputs when it comes time to execute, you need to strike out huge amounts of data. And, if you do that, you no longer have what you need to train the machine. Machines are bad at learning from small amounts of data; nobody has figured out how to get a machine to learn as a human infant can from a single experience. Nor has anybody, at least yet, found a way to take a scalpel to datasets; there's no way, in the state of the art, to excise "bad" data reliably for purposes of training a machine.[20] Accordingly, data pruning is anathema to computer scientists.[21]

As for legal proceedings, data pruning is, as we said, a complete answer to the problem it addresses—in situations in which the data was pruned before a decision-maker sees it. As we noted, however, not all improper evidence stays out of the court room. Nor does all knowledge gained from improper evidence—fruit of poisonous trees—stay out. Once it enters, which is to say once a decision-maker, such as a juror, has learned it, its potential for mischief is there. You cannot undo facts. They exist. Experience is a fact. Things that have been experienced, knowledge that has been gained, do not disappear by fiat.

A formalist would posit that the only facts that affect the trial process are those that the filters of evidentiary exclusion are designed to let in. As we have discussed, however, Holmes understood the law, including the results of trials, to derive from considerably more diverse material. Juries, lawyers, and judges all come with their experiences and their prejudices. To Holmes, these were a given, which is why he thought trying to compel decision-makers "to testify to the operations of their minds in doing the work entrusted to them" was an "anomalous course" and fruitless.[22] You cannot simply excise the unwanted experience from someone's mind—any more than present-day computer scientists have succeeded in cutting the "bad" data from the training dataset.

8.3 INFERENTIAL RESTRAINT

What you *can* do—however imperfect a strategy it may be—is place limits on what you allow yourself, the jury, the machine, or the judge to *infer* from the data or the experience. Inferential restraint is familiar in both law and machine learning. In efforts to address the problem of bad evidence (bad data) in machine learning, most of the energy indeed has been directed toward this approach: instead of pruning the data, computer scientists are developing methods to restrict the type of inferential outputs that the machine is able to generate.[23]

In the legal setting, placing restrictions upon inferences has been an important strategy for a long time. Judges' instructions to juries serve that purpose; appeals courts recognize that judges' instructions, properly given, have curative effect.[24] Judges, in giving curative instructions, understand that, even when bad evidence of the kind addressed in *Silverthorne* and *Nardone* (evidence seized in violation of a constitutional right) has been stopped before it gets to the jury, there still might be knowledge in the jurors' minds that could exercise impermissible effects on their decision. The jurors might have gained such knowledge from a flip word in a lawyer's closing argument.[25] They might have brought it with them in off the street in the form of their life experiences; Holmes understood juries to have a predilection for doing just that.[26] Knowledge exists which is to be kept from affecting verdicts, if those verdicts are to be accepted as sound. But some knowledge comes to light too late to prune. There, instead, a cure is to be applied. In the courtroom, the cure takes the form of an instruction from the judge. The instruction tells the jurors to restrain the inferences they draw from certain evidence they have heard. The restraint is intended to operate in the mental machinery of each juror.

A further situation that calls for inferential restraint is that in which some piece of evidence has probative value and may be used for a permissible purpose, but a risk exists that a decision-maker might use the evidence for an impermissible purpose. Pruning the evidence would have a cost: it would entail losing the probative value. Thus, as judges tell jurors to ignore certain experiences that they bring to the court room and certain bad evidence or statements that, despite best efforts, have entered the court room, so do judges guide jurors in the use of knowledge that the court deliberately keeps.[27] Here, too, analogous approaches are being explored in machine learning.[28]

8.4   EXECUTIONAL RESTRAINT

From Holmes's judgment in *Silverthorne*, one discerns that a strategy of restraint operates not just on the mental processes of the people involved at a given time but also on their future conduct and decisions. *Silverthorne* was a statement to the government about how it was to use knowledge. True, the immediate concern was to cut out the bad evidence root and branch, to keep it from undermining judicial procedure and breaching a party's constitutional rights. Data pruning is what generations of readers of *Silverthorne* understand it to have done; the principle of the fruit of the poisonous tree more widely indeed is read as a call for getting rid of problematic inputs.[29]

There is more to the principle of the fruit of the poisonous tree, however, than data pruning. Consider closely what Holmes said in *Silverthorne*: "the knowledge *gained* by the Government's own wrong cannot be *used* by it in the way proposed" (emphasis added). So the "Government's own wrong" already had led it to gain certain knowledge. Holmes was not proposing the impossible operation of cutting that knowledge from the government's mind. The time for pruning had come and gone. Holmes was proposing, instead, to restrain the Government from executing future actions that the Government on the basis of that knowledge might otherwise have executed: knowledge gained by the Government's wrong was not to be "used by it." The poisonous tree (to use Frankfurter's expression) addresses a state of the world *after* the bad evidence has already generated knowledge. The effect of that knowledge on future conduct is what is to be limited. That is to say, executional restraint, the strategy of restricting what action it is permissible to execute, inheres in the principle.[30]

8.5   POISONOUS PASTS AND FUTURE GROWTH

Seen in this, its full sense, the principle of the fruit of the poisonous tree has high salience for machine learning, in particular as people seek to use machine learning to achieve outcomes society desires. A training dataset necessarily reflects a past state of affairs.[31] The future will be different. Indeed, in many ways, we *desire* the future to be different, and we work toward making it so in particular, desirable ways. But change, as such, doesn't require our intervention. Even if we separate ourselves from our desires for the future, from values that we wish to see reflected in the society of tomorrow, it is a matter of empirical observation, a fact, that the future *will* be different. Thus, either way, whether or not our values enter into it, we err if we rely blindly on a mechanism whose outputs are a faithful reflection of the inputs from the past that shaped it. We must therefore restrain the conclusions that we draw from those outputs, and the actions we take, or else we will be getting the future wrong.

In machine learning, there is widespread concern about undesirable correlations. An example could be supplied by a machine that hands out prison sentences. The machine is based on data. The data is a given. Americans of African ancestry have received a disproportionate number of prison sentences. Trained on that data, a machine will give reliable results: it will give results that reliably install the past state of affairs onto its future outputs. African-Americans will keep getting a disproportionate number of prison sentences. Reliability here has no moral valence in itself; it connotes no right or wrong. It is simply a property of the machine. The reason society objects to reliability of this kind, when considering an example as obvious as the prison sentencing machine, is that this reliability owes to data collected under conditions that society hopes will not pertain in the future. We want to live under new conditions. We do not want a machine that perpetuates the correlations found in *that* data and thus perpetuates (if we obey the machine) the old conditions. Some computer scientists think there may be ways to address this concern about undesirable correlations by pruning the training dataset.[32] We mentioned the technical challenges this presents for machine learning. We speculate that the other strategies will be as important in machine learning as they have been in law: restrain the inferences and actions that derogate the values we wish to protect. That's how we increase the chances that we'll get the future right.

NOTES

1. See in particular Chapters 6 and 7, pp. *67 ff*, *81 ff*.
2. Silverthorne Lumber Co. et al. v. United States, 251 U.S. 385, 392 (1920, Holmes, J.); Nardone et al. v. United States, 308 U.S. 338, 342 (1939, Frankfurter, J.).
3. For an exposition of the concept of probative value by reference to principles of probability, see Friedman, *A Close Look at Probative Value*, 66 B.U. L. Rev. 733 (1986).
4. Op. cit.
5. 251 U.S. at 390.
6. Id. at 392.
7. Id.
8. Id.
9. See Pitler, *"The Fruit of the Poisonous Tree" Revisited and Shepardized*, 56 Cal. L. Rev. 579, 581 (1968). Justice Frankfurter called it evidence

"used derivatively": Nardone et al. v. United States, 308 U.S. 338, 341 (1939, Frankfurter, J.). Cf. noting that "[t]he exclusionary prohibition extends as well to the indirect as the direct products of such invasions [of a premises in breach of constitutional right]": Wong Sun v. United States, 371 U.S. 471, 484 (1963) (Brennan, J.). See further Brown (Gen. Ed.), McCORMICK ON EVIDENCE (2006) § 176 pp. 292–94.

10. 308 U.S. at 342.

11. Id. at 341. As to the sufficiency of connection, see Kerr, *Good Faith, New Law, and the Scope of the Exclusionary Rule*, 99 GEO. L. J. 1077, 1099–1100 (2011). Cf. Devon W. Carbado, *From Stopping Black People to Killing Black People: The Fourth Amendment Pathways to Police Violence*, 105 CAL. L. REV. 125, 133–35 (2016).

12. Undesirable outcomes from a machine learning process are shot through with questions of causation—e.g., is it appropriate to hold accountable the computer scientist who engineered a machine learning system, when an undesirable outcome is traceable back to her conduct if at all then only by the most attenuated lines? Regarding the implications for tort law, see, e.g., Gifford, *Technological Triggers to Tort Revolutions: Steam Locomotives, Autonomous Vehicles, and Accident Compensation*, 11 J. TORT LAW 71, 143 (2018); Haertlein, *An Alternative Liability System for Autonomous Aircraft*, 31 AIR & SPACE L. 1, 21 (2018); Scherer, *Regulating Artificial Intelligence Systems: Risks, Challenges, Competences, and Strategies*, 29 HARV. J. L. TECH. 353, 363–366 (2016); Calo, *Open Robotics*, 70 MD. L. REV. 571, 602 (2011). Writers have addressed causation problems as well in connection with international legal responsibility and autonomous weapons: see, e.g., Burri, *International Law and Artificial Intelligence*, 60 GYIL 91, 101–103 (2017); Sassóli, *Autonomous Weapons and International Humanitarian Law: Advantages, Open Technical Questions and legal Issues to Be Clarified*, 90 INT'L L. STUD. 308, 329–330 (2014).

13. In the computer science literature, the expression "data pruning" has been associated with cleaning noisy datasets in order to improve performance. See, e.g., Anelia Angelova, Yaser S. Abu-Mostafa & Pietro Perona, *Pruning Training Sets for Learning of Object Categories*: CVPR Conference (2005), San Diego, June 20–25, 2005: vol. 1 IEEE 494–501.

14. Mapp v. Ohio, 367 U.S. 643, 656 (1961).

15. See for example Meriani, *Digital Platforms and the Spectrum of Data Protection in Competition Law Analysis*, 38(2) EUR. COMPET. L. REV. 89, 94–95 (2017); Quelle, *Enhancing Compliance Under the General Data Protection Regulation: The Risky Upshot of Accountability- and Risked-Based Approach*, 9 EUR. J. RISK REGUL. 502, 524–525 (2018).

16. "Bad evidence" is thus of broadly two types. (i) Evidence may be bad because the manner of its collection is undesirable. That type of bad evidence might have raised no problem, if its collection had not been

tainted. (ii) The other type is bad, irrespective of how the evidence collector behaved. It is bad, because it poses the risk of an invidious influence on the decision process itself.

17. Doctrinal writers on evidence have struggled to articulate how to determine whether the link between bad evidence and challenged evidence is attenuated enough to "dissipate the taint." Clear enough is the existence of an exception to the fruit of the poisonous tree. Unclear is when the exception applies. Here the main treatise on American rules of evidence has a go at an answer:

> This exception... does not rest on the lack of an actual causal link between the original illegality and the obtaining of the challenged evidence. Rather, the exception is triggered by a demonstration that the nature of that causal link is such that the impact of the original illegality upon the obtaining of the evidence is sufficiently minimal that exclusion is not required despite the causal link. Brown (Gen. Ed.), MCCORMICK ON EVIDENCE (2006) § 179 p. 297.

Note the circularity: the exclusion "exception is triggered" (i.e., the exclusion is not required) when the "exclusion is not required." The hard question is what precisely are the characteristics that give a causal link such a "nature" that it is "sufficiently minimal."

18. Dunn v. United States, 307 F.2d 883, 886 (Gewin, J., 5th Cir., 1962). Courts outside the U.S. have used the phrase too: Kung v. Peak Potentials Training Inc., 2009 BCHRT 154, 2009 CarswellBC 1147 para 11 (British Columbia Human Rights Tribunal, Apr. 23, 2009).

19. See for example Fuery et al. v. City of Chicago, 900 F. 3d 450, 457 (Rovner, J., 7th Cir., 2018).

20. Broadly speaking, there are two ways to prune a dataset: removing *items* from the dataset (rows) for example to remedy problems of unbalanced representation, or removing a sensitive *attribute* from the dataset (a column). It has been widely observed that removing a sensitive attribute is no use, if that attribute may be more or less reliably predicted from the remaining attributes. Removing items is also tricky: for example, the curators of the ImageNet dataset, originally published in 2009 (see Prologue, p. xiii, n. 23) were as of 2020 still playing whack-a-mole to remedy issues of fairness and representation. See Yang et al. (2019).

21. Of course we don't mean that computer scientists find the *goal* that motivates data pruning efforts to be antithetical morally or ethically. Instead, data pruning, a blunt instrument perhaps acceptable as a stop-gap, is at odds with how machine learning works.

22. Coulter et al. v. Louisville & Nashville Railroad Company, 25 S.Ct. at 345, 196 U.S. at 610 (1905).

23. For a cutting-edge illustration, see Madras et al. (2018). What is particularly interesting about their approach is that, in order to guarantee that the machine learning system's inferences are unbiased against individuals with some protected attribute x, that attribute *must be available* to the machine. This illuminates why computer scientists are uneasy about data pruning.

24. See Leonard (ed.), NEW WIGMORE (2010) § 1.11.5 p. 95 and see id. 95–96 n. 57 for judicial comment. The standard for establishing that limiting instructions have *failed* is exacting. See, e.g., Encana Oil & Gas (USA) Inc. v. Zaremba Family Farms, Inc. et al., 736 Fed. Appx. 557, 568 (Thapar, J., 6th Cir., 2018).

25. A problem addressed repeatedly by U.S. courts. See, e.g., Dunn v. United States, 307 F.2d 883, 885–86 (Gewin, J., 5th Cir. 1962); McWhorter v. Birmingham, 906 F.2d 674, 677 (Per Curiam, 11th Cir. 1990). A substantial literature addresses jury instructions, including from empirical angles. See, e.g., Mehta Sood, *Applying Empirical Psychology to Inform Courtroom Adjudication—Potential Contributions and Challenges*, 130 HARV. L. REV. F. 301 (2017).

26. See Chapter 7, p. 81 *ff.* See also Liska, *Experts in the Jury Room: When Personal Experience is Extraneous Information*, 69 STAN. L. REV. 911 (2017).

27. Appeals courts consider such instructions frequently. For a recent example, see United States v. Valois, slip. Op. pp. 13–14 (Hull, J., 2019, 11th Cir.). Cf. Namet v. U.S., 373 U.S. 179, 190, 83 S.Ct 1151, 1156 n. 10 (Stewart, J., 1963).

28. See Madras et al., op. cit.

29. That reading is seen in judgments, including (perhaps particularly) of foreign courts when they observe that "fruit of the poisonous tree" is not part of their law. See, e.g., Z. (Z.) v. Shafro, 2016 ONSC 6412, 2016 CarswellOnt 16284, para 35 (Kristjanson, J., Ontario Superior Court of Justice, Oct. 14, 2016). Some foreign courts do treat the doctrine as part of their law and apply a similar reading. See, e.g., Dela Cruz v. People of the Philippines (2016) PHSC 182 (Leonnen, J., Philippines Supreme Court, 2016), with precedents cited at Section III, n. 105. See the comparative law treatment of the principle by Thaman, *"Fruits of the Poisonous Tree" in Comparative Law*, 16 SW. J. INT'L L. 333 (2010).

30. Executional restraint and inferential restraint, as we stipulate the concepts, in some instances overlap, because an execution that is to be restrained might be a mental or computational process of inference. Overlap is detectable in *Silverthorne Lumber*. The Government, Holmes said, was to be restrained from how it used the knowledge that it had gained through an illegal search and seizure. The use from which Holmes called the Government to be restrained was equally the Government's *reasoning* about where to go in search of evidence; and the *physical action* it executes in the

field. Restraint has both aspects as well where one is concerned, instead of with preventing people from using knowledge to generate more knowledge, with preventing a machine learning system from using an input to generate more outputs. The overlap arises in machine learning between the two variants of restraint, because machine learning systems (at least in the current state of the art) don't carrying on computing with new inputs unless some action is taken to get them to execute. The executional restraint would be to refrain from switching on the machine (or, if its default position is "on," then to switch the machine off).

The overlap is also significant where human institutions function under procedures that control who gets what information and for what purposes. Let us assume that there is an institution that generates decisions with a corporate identity—i.e., decisions that are attributable to the institution, rather than to any one human being belonging to it. Corporations and governments are like that. Let us also assume that, in order to generate a decision that bears the corporate identity, two or more human beings must handle certain information; and one of them, or some third person, has the power to withhold that information. The person having the withholding power may place a restraint upon the institution: she may withhold the information and, thus, the institution cannot carry out the decision process. The restraint in this setting has overlapping characteristics. It is inferential, in that it restrains the decision process; it is executional, in that it restrains the actions of the individual constituents of the institution.

31. See Chapter 3, p. 37.
32. Chouldechova & Roth, op. cit., Section 3.4 p. 7. Cf. Paul Teich, *Artificial Intelligence Can Reinforce Bias*, FORBES (Sept. 24, 2018) (referring to experts who "say AI fairness is a dataset issue").

CHAPTER 9

From Holmes to AlphaGo

Holmes in *The Path of the Law* asked "What constitutes the law?" and answered that law is nothing more than prophecies of what the courts will do. As we discussed in Chapter 5, this is not just the trivial observation that one of the jobs of a lawyer is to predict the outcome of a client's case: it is the insight that growth and development of the law itself—the *path* of the law—is constituted through predictive acts.

Holmes was preoccupied throughout his legal career with understanding the law as an evolving system. Kellogg, in a recent study of the roots of Holmes's thinking,[1] traces this interest to the period 1866–1870, Holmes's first years as a practicing lawyer, and to his reading of John Stewart Mill on the philosophy of induction, and of William Whewell and John Herschel on the role of induction in scientific theory-building. Holmes's original insight was that the development of law is a *process of social* induction: it is not simply logical deduction from axioms laid down in statutes and doctrine as formalists would have it; it is not simply the totality of what judges have done as the realists would have it. Instead, law develops through agents *embedded in society* who take actions that depend on and contribute to the accumulating body of experience, and it involves *social* agents who through debate are able to converge toward entrenched legal doctrine.

In the standard paradigm for machine learning, there is no counterpart to the first part of Holmes's insight of social induction—i.e., to the role of active agents embedded in society. The standard paradigm is that there is something for the machine to learn, and this "something" is data, i.e.

© The Author(s) 2020
T. D. Grant and D. J. Wischik, *On the path to AI*,
https://doi.org/10.1007/978-3-030-43582-0_9

given, and data does not accumulate through ongoing actions. This is why the field is called "machine *learning*" rather than "machine *doing*"! Even systems in which a learning agent's actions affect its surroundings, for example a self-driving car whose movements will make other road-users react, the premise is that there are learnable patterns about how others behave, and learning those patterns is the goal of training, and training should happen in the factory rather than on the street.

There is however a subfield of machine learning, called reinforcement learning, in which the active accumulation of data plays a major role. "AlphaGo,"[2] the AI created by DeepMind which in 2016 won a historic victory against top-ranking (human) Go player Lee Seedol, is a product of reinforcement learning. In this chapter we will describe the links between reinforcement learning and Holmes's insight that law develops through the actions of agents embedded in society.

The second part of Holmes's insight concerns the process whereby data turns into doctrine, the "continuum of inquiry."[3] As case law accumulates, there emerge clusters of similar cases, and legal scholars, examining these clusters, hypothesize general principles. Holmes famously said that "general propositions do not decide concrete cases," but he also saw law as the repository of the "ideals of society [that] have been strong enough to reach that final form of expression." In other words, legal doctrine is like an accepted scientific theory[4]: it provides a coherent narrative, and its authority comes not from prescriptive axioms but rather from its ability to explain empirical data. Well-settled legal doctrine arises through a social process: it "embodies the work of many minds, and has been tested in form as well as substance by trained critics whose practical interest is to resist it at every step."[5]

There is nothing in machine learning that corresponds to this second aspect of Holmes's social induction, to the social dialectic whereby explanations are generated and contested and some explanation eventually becomes entrenched. In the last part of this chapter we will discuss the role of legal explanation, and outline some problems with explainability in machine learning, and suggest how machine learning might learn from Holmes.

9.1   Accumulating Experience

According to Holmes, "The growth of the law is very apt to take place in this way: two widely different cases suggest a general distinction, which is a clear one when stated broadly. But as new cases cluster around the opposite poles, and begin to approach each other [...] at last a mathematical line is arrived at by the contact of contrary decisions."[6]

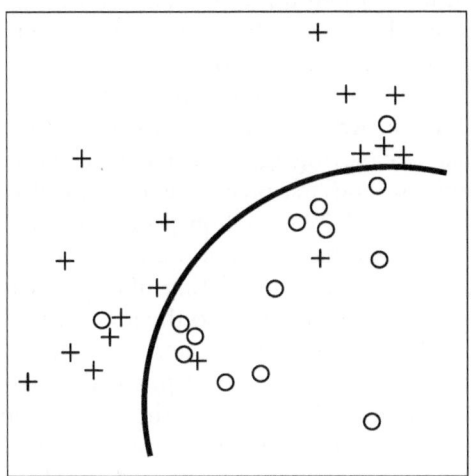

Holmes's metaphor, of a mathematical line drawn between cases with contrary decisions, will be very familiar to students of machine learning, since almost any introductory textbook describes machine-learning classification using illustrations such as the figure above. In the figure, each datapoint is assigned a mark according to its ground-truth label,[7] and the goal of training a classifier is to discover a dividing line. DeepMind's AlphaGo can be seen as a classifier: it is a system for classifying game-board states according to which move will give the player the highest chance of winning. During training, the system is shown many game-board states, each annotated according to which player eventually wins the game, and the goal of training is to learn dividing lines.

Holmes was interested not just in the dividing lines but in the accumulation of new cases. Some new cases are just replays with variations in the facts, Cain killing Abel again and again through history. But Holmes had in mind new cases arising from novel situations, where legal doctrine has not yet drawn a clear line. The law grows through a succession of particular legal disputes, and in no situation would there be a meaningful legal dispute if the dividing line were clear. Actors in the legal system adapt their actions based on the body of legal decisions that has accumulated, and this adaptation thus affects which new disputes arise. New disputes will continue to arise to fill out the space of possible cases, until eventually it becomes possible to draw a line "at the contact of contrary decisions." Kellogg summarizes Holmes's thinking thus: "he reconceived logical induction as a social process, a form of inference that engages adaptive action and implies social transformation."[8]

Machine learning also has an equivalent of adaptive action. The training dataset for AlphaGo was not given a priori: it was generated during training, by the machine playing against itself. To be precise, AlphaGo was trained in three phases. The first phase was traditional machine learning, from an a priori dataset of 29.4 million positions from 160,000 games played by human professionals. In the second phase, the machine was refined by playing against an accumulating library of earlier iterations of itself, each play adding a new iteration to the library. The final iteration of the second-phase machine was played against itself to create a new dataset of 30 million matches, and in the third phase this dataset was used as training data for a classifier (that is to say, the machine in the third phase trains on a given dataset, which, like the given dataset in the first phase, is not augmented during training). The trained classifier was the basis for the final AlphaGo system. DeepMind later created an improved version, AlphaGo Zero,[9] which essentially only needed the second phase of training, and which outperformed AlphaGo. The key feature of reinforcement learning, seen in both versions, is that the machine is made to take actions during training, based on what it has learnt so far, and the outcomes of these actions are used to train it further—Kellogg's "adaptive action."

Holmes says that the mathematical line is arrived at "by the contact of contrary decisions." Similarly, AlphaGo needed to be shown sufficient diversity of game-board states to fill out the map, so that it can learn to classify any state that it might plausibly come across during play. In law the new cases arise through fractiousness and conflict—"man's destiny is to fight"[10]—whereas for AlphaGo the map was filled out by artificially adding noise to the game-play dataset.

Holmes has been criticized for putting forwards a value-free model of the law—he famously defined truth "as the majority vote of that nation that can lick all the others."[11] Kellogg absolves Holmes of this charge: he argues that Holmes saw law as a process of social inquiry, using the mechanism of legal disputes to figure out how society works, similar to how science uses experiments to figure out how nature works. The dividing lines that the law draws are therefore not arbitrary: "Any successful conclusions of social inquiry must, in an important respect, conform with the world at large. Social inductivism does not imply that the procedures and ends of justification are relativist products of differing conventions."[12] Likewise, even though the training of AlphaGo is superficially relativist (it was trained to classify game-board states by the best next move, *assuming that its opponent is AlphaGo*), it is nonetheless validated by objective game mechanics: pitted against Lee Seedol, one of the top human Go players in the world, AlphaGo won.

9.2   Legal Explanations, Decisions, and Predictions

"It is the merit of the common law," Holmes wrote, "that it decides the case first and determines the principle afterwards."[13] Machine learning has excelled (and outdone the ingenuity of human engineers) at making decisions, once decision-making is recast as a prediction problem as described in Chapter 5. This success, however, has come at the expense of explainability. Can we learn how to explain machine learning decisions, by studying how common law is able to determine the principle behind a legal decision?

In the law, there is a surfeit of explanation. Holmes disentangled three types: (i) the realist explanation of why a judge came to a particular decision, e.g. because of an inarticulate major premise, (ii) the formalist explanation that the judge articulates in the decision, and (iii) explanation in terms of principles. Once principles are entrenched then the three types of explanation will tend to coincide, but in the early stages of the law they often do not. Principles reflect settled legal doctrine that "embodies the work of many minds and has been tested in form as well as substance by trained critics whose practical interest is to resist it at every step." They arise through a process of social induction, driven forwards not just by new cases (data) but also by contested explanations.

To understand where principles come from, we therefore turn to judicial decisions. (In legal terminology, *decision* is used loosely[14] to refer both to the judgement and to the judge's explanation of the judgement.)

Here is a simple thought experiment. Consider two judges A and B. Judge A writes decisions that are models of clear legal reasoning. She takes tangled cases, cases so thorny that hardly any lawyer can predict the outcome, and she is so wise and articulate that her judgments become widely relied upon by other judges. Judge B on the other hand writes garbled decisions. Eventually a canny lawyer realizes that this judge finds in favor of the defendant after lunch, and in favor of the plaintiff at other times of day (her full stomach is the inarticulate major premise). Judge B is very predictable, but her judgments are rarely cited and often overturned on appeal.

If we think of law purely as a task of predicting the outcome of the next case, then judgments by A and by B are equivalent: they are grist for the learning mill, data to be mined. For this task, the quality of their reasoning is irrelevant. It is only when we look at the *development* of the legal system that reasoning becomes significant. Judge A has more impact on future cases, because of her clear explanations. "[T]he epoch-making ideas," Holmes wrote, "have come not from the poets but from the philosophers, the jurists, the mathematicians, the physicists, the doctors—from the men who explain, not from the men who feel."[15]

Our simple thought experiment might seem to suggest that it is reasoning, not prediction, that matters for the growth of the law. What then of Holmes's famous aphorism, that prophecy is what constitutes the law? Alex Kozinski, a U.S. Court of Appeals judge who thought the whole idea of inarticulate major premise was overblown, described how judges write their decisions in anticipation of review:

> If you're a district judge, your decisions are subject to review by three judges of the court of appeals. If you are a circuit judge, you have to persuade at least one other colleague, preferably two, to join your opinion. Even then, litigants petition for rehearing and en banc review with annoying regularity. Your shortcuts, errors and oversights are mercilessly paraded before the entire court and, often enough, someone will call for an en banc vote. If you survive that, judges who strongly disagree with your approach will file a dissent from the denial of en banc rehearing. If powerful enough, or if joined by enough judges, it will make your opinion subject to close scrutiny by the Supreme Court, vastly increasing the chances that certiorari will be granted. Even Supreme Court Justices are subject to the constraints of colleagues and the judgments of a later Court.[16]

Thus judges, when they come to write a decision, are *predicting* how future judges (and academics, and agents of public power, and public opinion) will respond to their decisions. Kozinski thus brings us back to prophecy and demonstrates the link with explanations "tested in form as well as substance by trained critics."

9.3   Gödel, Turing, and Holmes

We have argued that the decision given by a judge is written in anticipation of how it will be read and acted upon by future judges. The better the judge's ability to predict, the more likely it is that this explanation will become part of settled legal doctrine. Thus judges play a double role in the growth of the law: they are actors who make predictions; and they are objects of prediction by other judges.

There is nothing in machine learning that is analogous, no system in which the machine is a predictor that anticipates future predictors. This self-referential property does however have an interesting link to classic algorithmic computer science. Alan Turing is well known in popular culture for his test for artificial intelligence.[17] Among computer scientists he is better known for inventing the Turing Machine, an abstract mathematical model of a computer that can be used to reason about the nature and limits of computation. He used this model to prove in 1936[18] that there is a task that is impossible to solve on any computer: the task of deciding whether a given algorithm will eventually terminate or whether it will get stuck in an infinite loop. This task is called the "Halting Problem." A key step in Turing's proof was to take an algorithm, i.e. a set of instructions that tell a computer what to do, and represent it as a string of symbols that can be treated as *data* and fed as input into another algorithm. Turing here was drawing on the work of Kurt Friedrich Gödel, who in 1930 developed the equivalent tool for reasoning about statements in mathematical logic. In this way, Gödel and later Turing were able to prove fundamental results about the limits of logic and of algorithms. They analyzed mathematics and computation as self-referential systems.

In Turing's work, an algorithm is seen as a set of instructions for processing data, and, simultaneously, as data which can itself be processed. Likewise, in the law, the judge is an agent who makes predictions, and, simultaneously, an object for prediction. Through these predictions, settled legal principles emerge; in this sense the law can be said to be constituted by prediction. Machine learning is also built upon prediction—but machine learning is not *constituted* by prediction in the way that law is. We might say that law is post-Turing while machine learning is still pre-Turing.[19]

9.4   WHAT MACHINE LEARNING CAN LEARN FROM HOLMES AND TURING

Our point in discussing legal explanation and self-referential systems is this:

(i) social induction in the law is able to produce settled legal principles, i.e. generally accepted explanations of judicial decision-making;

(ii) the engine for social induction in the law is prediction in a self-referential system;

(iii) machine learning has excelled (and outdone human engineering ingenuity) at predictive tasks for which there is an empirical measure of success;

(iv) if we can combine self-reference with a quantitative predictive task, we might get explainable machine learning decisions.

In the legal system, the quality of a decision can be evaluated by measuring how much it is relied on in future cases, and this quality is intrinsically linked to explanations. Explanations are evaluated not by "are you happy with what you've been told?", but by empirical consequences. Perhaps this idea can be transposed to machine learning, in particular to reinforcement learning problems, to provide a metric for the quality of a prediction. This would give an empirical measure of success, so that the tools that power machine learning can be unleashed, and "explainability" will become a technical challenge rather than a vague and disputed laundry list. Perhaps, as in law, the highest quality machine learning systems will be those that can internalize the behavior of other machines. Machines that do *that* would all the more trace a path like that of Holmes's law.

These are speculative directions for future machine learning research, which may or may not bear fruit. Nonetheless, it is fascinating that Holmes's understanding of the law suggests such avenues for research in machine learning.

NOTES

1. Kellogg (2018) 29.
2. Silver et al. (2016).
3. Kellogg (2018) 8.
4. Kellogg draws out Holmes's formative exposure to philosophers of science and his program to find a legal analogy for scientific hypothesis-formulation. Id. 25, 51.

5. Holmes, *Codes, and the Arrangement of the Law*, AMERICAN LAW REVIEW 5 (October 1870): 1, reprinted in Kellogg (1984) 77; *CW* 1:212. See Kellogg (2018) 37.
6. Holmes, *The Theory of Torts*, AMERICAN LAW REVIEW 7 (July 1873): 652, 654. See Kellogg (2018) 41.
7. See Chapter 3, p. 37.
8. Kellogg (2018) 17.
9. Silver et al. (2017).
10. Holmes to Learned Hand, June 24, 1918, quoted in Kellogg (2018) 186–87.
11. Id.
12. Kellogg (2018) 180.
13. *Codes, and the Arrangement of the Law*, AMERICAN LAW REVIEW 5 (October 1870): 1, reprinted in Kellogg (1984) 77; *CW* 1:212. See Kellogg (2018) 37.
14. In Chapter 6 we drew attention to a similar confounding, in Hempel's failure to distinguish between explanation and prediction. See p. 74.
15. *Remarks at a Tavern Club Dinner (on Dr. S. Weir Mitchell)* (March 4, 1900) reprinted in De Wolfe Howe (ed.) (1962) 120. Poetry was one of Holmes's polymath father's several avocations.
16. Kozinski, *What I Ate for Breakfast and Other Mysteries of Judicial Decision Making*, 26 LOY. L.A. L. REV. 993 (1993). Kozinski (1950–) served on the U.S. Court of Appeals for the Ninth Circuit from 1985 to 2017.
17. See *The Imitation Game* (U.S. release date: Dec. 25, 2014), in which actor Benedict Cumberbatch plays Turing. For the test, see Alan M. Turing, *Computing Machinery and Intelligence*, 59 MIND 433–60 (1950) (and esp. the statement of the game at 433–34). Cf. Halpern, *The Trouble with the Turing Test*, THE NEW ATLANTIS (Winter 2006): https://www.thenewatlantis.com/publications/the-trouble-with-the-turing-test.
18. Alan Turing, *On Computable Numbers, with an Application to the Entscheidungsproblem*, 2(42) LMS PROC. (1936). For a readable outline of Turing's contribution and its historical context, see Neil Immerman, *Computability and Complexity*, in Zalta (ed.), THE STANFORD ENCYCLOPEDIA OF PHILOSOPHY (Winter 2018 edition): https://plato.stanford.edu/archives/win2018/entries/computability/.
19. For a further link between Turing and Holmes, see Chapter 10, p. 123.

Conclusion

In every department of knowledge, what wonderful things have been done to solicit the interest and to stir the hopes of new inquirers! Every thing is interesting when you understand it, when you see its connection with other things, and, in the end, with all other things.

> *Oliver Wendell Holmes, Jr.,* Remarks at the Harvard Commencement Dinner *(June 22, 1880)*[1]

This book has posited a connection between two seemingly remote things: between a late nineteenth century revolution in law and an early twenty-first century revolution in computing. The jurist on whose work we've drawn, if he'd been transported to the present day, we think would have been open to the connection. Holmes's early milieu had been one of science, medicine, and letters, these being fields in which his father had held a prominent place and in which their city, in Holmes's youth, in America had held the preeminent place.[2] Leading lights of nineteenth century philosophy and science numbered among Holmes's friends and interlocutors at home and abroad in the years immediately after the Civil War. Holmes continued throughout his life to engage with people whom today we would call technologists. His interest in statistics and in the natural sciences was broad and deep and visible in Holmes's vast output as a scholar and a judge. Lawyering and judging, to Holmes, were jobs but also objects to be searched for deeper understanding.

© The Author(s) 2020
T. D. Grant and D. J. Wischik, *On the path to AI*,
https://doi.org/10.1007/978-3-030-43582-0_10

We hope that in the preceding chapters, by considering "its connection with other things," we have contributed to a deeper understanding of where machine learning belongs in the wider currents of modern thought. The common current that has shaped both legal thought and computer science is probability. It is a strikingly modern concept. As we have recalled, its origins are not less recent than the mid-seventeenth century. Its impact has been felt in one field after another, though hardly all at once. Law was present at its origins, though it took over two centuries before a new jurisprudence would take shape under its influence. Computing, too, did not begin as an operation in probability and statistics, but now probability and statistics are the indispensable core of machine learning. Thus both law and computing have undergone a shift from their earlier grounding in deductive logic: they have taken an inductive turn, based on pattern finding and prediction.

But, to conclude, let us turn away from intellectual history and look instead to the future.

10.1   HOLMES AS FUTURIST

Holmes, notwithstanding the strains of fatalism evident in his words, was fascinated by the potential for change, in particular change as driven by science and technology. Speaking in 1895 in honor of C. C. Langdell, that leading expositor of legal formalism, Holmes stated with moderate confidence that a march was on toward a scientific basis for law and that it would continue: "The Italians have begun to work upon the notion that the foundations of the law ought to be scientific, and, if our civilization does not collapse, I feel pretty sure that the regiment or division that follows us will carry that flag."[3] With the reference to "[t]he Italians" Holmes seems to have had in mind the positivism that was prevalent in legal theory in late nineteenth century Italy[4]; to the possibility of civilizational collapse, the pessimism prevalent generally in European philosophy at the time. In 1897, in *Law in Science and Science in Law*, Holmes hedged his prediction, but he continued to see contemporary advances in science and technology as pertinent to the organization of public life in the widest sense: "Very likely it may be that with all the help that statistics and every modern appliance can bring us there never will be a commonwealth in which science is everywhere supreme."[5] To entertain the possibility of a technological supremacy arising over law, even if to doubt that a scientific revolution in law would ever be complete, was still to place the matter in high relief.

Technological change had affected society at large for generations by the time Holmes wrote *Law in Science and Science in Law*. However, advances were accelerating and, moreover, in specific technical domains technology was interweaving itself with public order in unprecedented ways. This was the decade in which the U.S. Census Bureau first used a punch card machine with electric circuits to process data. The machine, known as the Hollerith Tabulator after its inventor, Herman Hollerith,[6] was the forerunner of modern data processing. Hollerith's company, the Tabulating Machine Company, was one of several later amalgamated to form the company that was eventually re-named IBM. The basic concept of the machine remained the cornerstone of data processing until the 1950s.[7] By 1911 (when Hollerith sold the Tabulating Machine Company), Hollerith Tabulators already had been used to process census data in the United States, United Kingdom, Norway, Denmark, Canada, and the Austrian and Russian Empires. Railroads, insurance companies, department stores, pharmaceutical companies, and manufacturers employed Hollerith machines as well.[8] The Hollerith Tabulator lowered the cost of handling large quantities of data and accelerated the work; the SCIENTIFIC AMERICAN, which ran an article on the machine in its August 30, 1890 edition, attributed the "early completion of the [census] count... to the improved appliances by which it was executed."[9]

The Hollerith machines did more than increase the efficiency of the performance of existing tasks, however. Because they enabled users to interrogate datasets in ways that earlier were prohibitively time-consuming—for example, asking how many people in the year 1900 in Cincinnati were male blacksmiths born in Italy—the Hollerith machines opened the door to new uses for data, not just more efficient head counts. The SCIENTIFIC AMERICAN referred to the "elasticity of function" that the machines enabled.[10] Hollerith himself was referred to as the first "statistical engineer."[11]

Holmes was not excited about putting his hands on the various innovations that technologists were bringing to the market; he doubted that his house would have had electricity or a telephone if his wife and not had them installed.[12] It would be surprising, however, if Holmes had not known of the Hollerith machine.[13] The edition of SCIENTIFIC AMERICAN containing the article about Hollerith and the census featured illustrations of the tabulator at work on its cover. The same periodical had run an article four years earlier on Holmes's father.[14] Holmes was a paid subscriber.[15] He also encountered technological innovations in the course of his principal employment: Holmes authored a number of the Supreme Court's decisions in patent matters[16] (none, it seems, concerning Hollerith, though the "statistical engineer" was no stranger to intellectual property disputes[17]). Curiously enough, Hollerith's headquarters and workshop were in a building in the Georgetown part of Washington, DC not many blocks from where Holmes lived after moving to the capital,[18] and the building in which the Census employed a large array of the machines was a short block off the most direct route (2.2 miles) between the Capitol (which then housed the Supreme Court) and Holmes's house at 1720 I Street, NW. Contemporaries remarked on the distinctive chimes that bells on the machines made, a noise which rose to a clamor in the building and which could be heard on the street below.[19] Holmes was a keen rambler whose peregrinations in Boston, Washington, and elsewhere took him in pursuit of interesting things.[20] Whether or not the Hollerith Tabulator was the appliance Holmes had in mind in *Law in Science and Science in Law*, technology was in the air. The emergence of modern bureaucracy in the early nineteenth century had been associated with an ambition to put public governance on a scientific basis;[21] the emergence of machines in the late nineteenth century that process data inspired new confidence that such an ambition was achievable.[22] To associate the commonwealth and its governance with statistics and "modern appliance," as Holmes did, was very much of a piece with the age.

Holmes's interest in technology induced him to maintain wide-ranging contacts, some of them rather idiosyncratic. A fringe figure named Franklin Ford[23] for a number of years corresponded with Holmes about the former's theories regarding news media and credit institutions. Ford imagined a centralized clearing mechanism that would give universal access to all news and credit information, an idea today weirdly evocative of the world wide web; and he said that this mechanism would supplant the state and its legal institutions, a prediction likewise evocative of futurists today who say, e.g., the blockchain will bring about the end of currencies issued under government fiat. Holmes continued the correspondence for years, telling Ford at one point that he (Ford) was "engaged with the large problems of the sociologist, by whom all social forces are equally to be considered and who, of course, may find and will find forces and necessities more potent than the theoretical omnipotence of the technical lawgiver."[24] Holmes evidently continued to speculate that law might in time give way to the experience embodied in "all social forces" which, in the context of that correspondence, suggested "social forces" mediated in some way by technology. In his correspondence with Franklin Ford, whose schemes aimed at the dissemination and use of data, Holmes seemed to intuit that, if machines came to martial data in even "more potent" ways, civilization-changing effects might follow.

Study of Justice and Mrs. Oliver Wendell Holmes's Washington, DC residence (Harris & Ewing, Washington, DC, United States [photographer] 1935; Harvard Law School Library, Historical & Special Collections; original at Library of Congress, Prints & Photographs Division, Lot 10304)

In a flight of fancy, in a 1913 speech, Holmes went so far as to specu-late about the evolution of the species:

> I think it not improbable that man, like the grub that prepares a chamber for the winged thing it never has seen but is to be—that man may have cosmic destinies that he does not understand... I was walking homeward on Pennsylvania Avenue near the Treasury, and as I looked beyond Sher-man's Statue to the west the sky was aflame with scarlet and crimson from the setting sun. But, like the note of downfall in Wagner's opera, below the sky line there came from little globes the pallid discord of the electric lights. And I thought to myself the *Götterdämmerung* will end, and from those globes clustered like evil eggs will come the new masters of the sky. It is like the time in which we live. But then I remembered the faith that I partly have expressed, faith in a universe not measured by our fears, a uni-verse that has thought and more than thought inside of it, and as I gazed, after the sunset and above the electric lights there shone the stars.[25]

Holmes in this passage holds his own with the most imaginative—and the most foreboding—twenty-first century transhumanists. The operatic ref-erence, with a little stretch, is even more evocative of change wrought by science than first appears. True, the characters in Wagner's opera don't use electric circuits for data processing.[26] But it is not too foreign to Holmes's speculations about the world-changing potential of statistics—or to the conceptual foundations of the machine learning age—that, in the Pro-logue to that last of the Ring Cycle operas, the Fates, whose vocation is to give prophecies, are weaving: and the rope with which they weave is made of the knowledge of all things past, present, and yet to come. The rope breaks, and thus the stage is set for the end of one world and the start of another.[27] Mythological data scientists foretelling the epochal changes their science will soon effect!

In less fanciful tenor, in *Law in Science* Holmes suggested that science might aid law and possibly replace it:

> I have had in mind an ultimate dependence upon science because it is finally for science to determine, so far as it can, the relative worth of our different social ends, and, as I have tried to hint, it is our estimate of the proportion between these, now often blind and unconscious, that leads us to insist upon and to enlarge the sphere of one [legal] principle and to allow another gradually to dwindle into atrophy.[28]

Science, in Holmes's view, would be put in harness to law; or it would replace law by taking over the social functions that law for the time being serves. Richard Posner, Chief Judge of the U.S. Court of Appeals for the Seventh Circuit at the time, on the 100th anniversary of *The Path of the Law* read Holmes to contemplate that law would be "succeeded at some time in the future by forms of social control that perform the essential functions of law but are not law in a recognizable sense."[29] The most accomplished scholar of Holmes to have served on an American court in the present century, Posner also thought the passage about *Götterdämmerung* and "cosmic destinies" noteworthy.[30] Whatever the precise role Holmes contemplated for science, and wherever he thought science would take us, it is evident that Holmes's philosophy did not equate with narrow presentism. Holmes was keenly interested in the future, including the future impact of science on law.

Writers have cautioned against "scientism,"[31] the unjustified confidence in the potential for science to solve society's problems. Our focus here has not been to repeat well-known critiques of unexamined enthusiasm for technological change. The acknowledgement of correlation between new technologies and risk has tempered scientistic impulses[32]; admonitions have been sounded in regard to Holmes's ideas about science.[33] It nevertheless is timely to alert practitioners of computer science that they ignore sanguinary lessons of the history of ideas if they place blind faith in the power of their craft. Holmes, perhaps, can be read for cautionary notes in that regard.

But our chief purpose in this book has been to use the analogy from Holmes's jurisprudence to cast light on machine learning. Let us ask, then, what, if any, lessons for the future of computer science might be found in Holmes's speculations about the future of law.

10.2 WHERE DID HOLMES THINK LAW WAS GOING, AND MIGHT COMPUTER SCIENCE FOLLOW?

Holmes, in thinking about law, found interest in wider currents that law both is borne upon and drives. Holmes considered the possibility that science will replace law—more precisely, that scientific method and technological advances will reveal rules and principles that law will adopt and thus give law a more reliable foundation. A curious irony would be if it went the other way around. A self-referential system—prediction as the system's output and its input as well—which is to say Holmes's concept of law as he thought law actually is—is what computer scientists, we speculate, might seek to make machine learning into. That has not been what machine learning is. True, machine learning has moved beyond logic and so is now an inductive process of finding patterns in input data to attain an output. So far however that is the end of the road. If machine learning goes further, if it comes to embody self-referential mechanisms such as Gödel and Turing devised in mathematics and computation, then machine learning will come to look even more like Holmes's law—an inductive system of prediction-making and self-referential prediction-shaping. The law, as Holmes understood it, would then have foreshadowed the future of computer science.

This is not how Holmes seems to have imagined things would go. We discern in his futurist and scientistic vein that Holmes thought that law, as he understood it to be, would give way to something else. What he thought law as prediction would give way to is not clear, but, as Judge Posner suggested, Holmes seems to have contemplated that science and technology would end society's reliance on law and bring about new mechanisms of control. The new mechanisms would be based on logic, rather than experience, and thus, in Holmes's apparent vision, would come full circle back to a sort of formalism—not a formalism based on arbitrary doctrines and rules, but based, instead, on propositions derived from what nineteenth century thinkers conceived of as science.

Holmes's speculations about science replacing law would seem to have a genealogy back to Leibniz, though we are not aware to what extent, if at all, Holmes was thinking about that antecedent when he wrote about a "scientific" future for law. Leibniz wrote about the possible use of mathematical models to describe law and philosophy in sufficient detail and scope that (in Leibniz's words), "if controversies were to arise, there would be no more need of disputation between two philosophers than between two accountants. For it would suffice for them to take their pencils in their hands and to sit down at the abacus and say to each other (with a friend if they wish): Let us calculate."[34] It is indeed this branch of Leibniz's thought that interests people, like Michael Livermore, who are considering how to put state of the art computing to work on legal problems.[35] As we have suggested, however, it is Leibniz's thinking about probability, not his speculation that fixed rules might one day answer legal questions, that has special resonance in a machine learning age. Leibniz thus, arguably, presaged Holmes, both in the application of probability theory to law and in the speculation that such application ultimately might be set aside in favor of a universal body of rules. What is more, he may have presaged Holmes, too, in thinking past that part of his thinking that has real salience to machine learning.

The irony, then, would be if computing followed Holmes's description of law as he thought law is—not his speculations about where he thought law was going. Machine learning today finds itself on the path that Holmes understood law actually to traverse in his own day. Machine learning has shifted computer science from logical deduction to an inductive process of finding patterns in large bodies of data. Holmes's realist conception of law shifted the law from rules-based formalism to a search for patterns in the collected experience of society. Reading Holmes as he understood the law to be, not his speculations about where law might go, we discern a path of the law that very much resembles that taken by machine learning so far.

Along that path, Holmes supplied a complete description of law. He described law as prophecy—meaning that all instances of law, all its expressions, are prophecy, and each successive prediction, whatever its formal source, in turn shapes, to a greater or to a lesser degree, all the prophecies to come. There is thus a self-referential character in law's inputs and outputs. In such self-reference, the law perhaps even anticipates a way ahead for machine learning: experience supplies the input from which present decisions are reached; and, in turn, those outputs become the inputs for future decisions. In short, though Holmes might have been waiting for technology to inform law, it could turn out that it is law that informs technology. The lawyers might have something to teach the computer scientists.

10.3   Lessons for Lawyers and Other Laypeople

Through a reading of Gödel, Turing, and Holmes in Chapter 9, we've identified a self-referential path that machine learning might follow. Regardless of where the technology goes from here, however, it is already too important for laypeople, including lawyers, to ignore. Thus we recall our initial task: to explain machine learning in terms that convey its essentials to the non-specialist.

We have aimed in the chapters above to convey the essentials. We have done so with the aid and within the limits of an analogy between two revolutions—one in jurisprudence, one in computing. While a much wider audience needs to come to grips with machine learning, lawyers, in view of their function in society, find themselves involved in distinctive ways in the questions it presents. Coming at machine learning through a legal analogy hopefully has established some new connections which will help non-specialists in general. The connections likely will have particular salience for the lawyers.

Lawyers whether by inclination or by habit are conservative. Law has much to do with authority, and legal argument seldom wins praise for conspicuous innovation. Legal minds, moreover, are skeptical; the enthusiasm for new machines that enlivens a technologist is not prevalent among lawyers. And, yet, lawyers from time to time have been involved in revolutions. As we noted at the start of this exploration of the conceptual foundations of machine learning, probability theory—the common current on which the two revolutions addressed in the chapters above have been carried—owes much to thinkers who were educated in law. So, too, long after, influential ideas in law have come from lawyers whom science and technology have interested.

H.L.A. Hart, having dedicated his Inaugural Lecture in 1952 as Professor of Jurisprudence at Oxford to *Definition and Theory in Jurisprudence*,[36] a few years later addressed his Holmes Lecture at Harvard to the challenge that arises when the legal system is called on to classify the sorts of "wonderful things" that solicited Holmes's interest time and again through his career. "Human invention and natural processes," Hart wrote...

> continually throw up such variants on the familiar, and if we are to say that these ranges of facts do or do not fall under existing rules, then the classifier must make a decision which is not dictated to him, for the facts and phenomena to which we fit our words and apply our rules are as it were *dumb*... Fact situations do not await us neatly labeled, creased, and folded, nor is their legal classification written on them to be simply read off by the judge.[37]

The impact of machine learning, realized and anticipated, identifies it as a phenomenon that Hart would have recognized as requiring legal classification. Lawyers and judges are called upon to address it with what rules they already have to hand. New legislation has attempted to address it in fresh terms. Explainability, an objective that we considered above, has motivated a range of new legislation, such as the GDPR, which entered into force in 2018 in the EU.[38] Enactments elsewhere pursue a similar objective, such as the California Consumer Privacy Act (CCPA) which enters into force in 2020,[39] as do a great many more.[40] It is both too early and beyond the scope of the present short book to take stock of the legislative output. It is not too early to observe the need for a wider, and more intuitive, understanding of what the legislator is being called upon to address. Law makers and practitioners need to understand the shift from algorithms to machine learning if they are to make good law and to practice it effectively. To attempt to label, crease, and fold machine learning into a familiar, algorithmic form is a mistake that Hart would have cautioned us to avoid.

Interestingly enough, Turing, so influential a figure in the line of human ingenuity that has interested us in the preceding chapters, seems to have been not too far removed from Holmes. The proximity was via Hart. Held by some the foremost legal philosopher since Holmes, Hart called Holmes a "heroic figure in jurisprudence."[41] Hart addressed the earlier jurist's idea of the law in detail, partly in riposte to Holmes's critics.[42] Hart did not write about Turing, but they were contemporaries—and linked. During World War II, which was before Hart embarked on a career as a legal academic, he was assigned to MI5, the British domestic intelligence agency. Hart's responsibility was to lead the liaison unit between MI5 and project ULTRA, the latter having been under the jurisdiction of MI6, the external intelligence agency. It was under project ULTRA, at a country house at Bletchley Park in England, that Turing did his codebreaking and developed the computational strategies that provided the point of departure for modern computing. Turing's work at Bletchley Park enabled MI6 to decipher encrypted German communications. So closely, however, did MI6 guard ULTRA that it was not clear at the start that the liaison unit for which Hart was responsible would serve any purpose. It appears that Hart's personal relations with key people in ULTRA played a role in getting the liaison to function—and, thus, in helping assure that Turing's technical achievements would add practical value to the war effort.[43] An eminent former student and colleague, John Finnis, notes that Hart never divulged further details about his wartime duties.[44] Years after the war—but still some time before Turing's rise to general renown—Hart did mention Turing: he mentioned to family that he admired him very much.[45]

That Turing's renown now extends well beyond computer science[46] evinces the wider recognition of computing's importance to modern society. Machine learning, as the branch of computing that now so influences the field, requires a commensurate breadth of understanding. We have written here about jurisprudence and the path to AI. Machine learning's impact, however, extends well beyond the legal profession. Every walk of life is likely to feel its impact in the years to come. Existing rules might help with some of the problems to which the new technology will give rise, but lawyers and judges will not find all the answers ready to "read off" the existing rules. We hope that having presented the ideas and ways of thinking behind machine learning through an analogy with jurisprudence will help lawyers to fold the new technology into the law—and will help laypeople fold it into the wider human experience across which machine learning's impact now is felt.

At the very least, we hope that lawmakers and people at large will stop using the word "algorithm" to describe machine learning, and that they will ask for "the story behind the training data" rather than "the logic behind the decision."

NOTES

1. The Occasional Speeches Of Justice Oliver Wendell Holmes, compiled by Mark De Wolfe Howe, Cambridge, MA: The Belknap Press of Harvard University Press, Copyright © 1962 by the President and Fellows of Harvard College. Copyright © renewed 1990 by Molly M. Adams.
2. It was Holmes's father who referred to Boston as the "hub of the universe," in part to mock its inhabitants' self-importance, but also to observe the real importance of the city for arts and sciences at that time: Budiansky 23–27.
3. *Learning and Science*, Speech at a Dinner of the Harvard Law School Association in Honor of Professor C.C. Langdell (June 25, 1895), De Wolfe Howe (ed.) (1962) 84, 85.
4. See for an overview Faralli, *Legal Philosophy in Italy in the Twentieth Century* in Pattaro & Roversi (eds.), A TREATISE OF LEGAL PHILOSOPHY AND GENERAL JURISPRUDENCE (2016) 369 *ff.*
5. 12 HARV. L. REV. at 462.
6. See https://www.census.gov/history/www/innovations/technology/the _hollerith_tabulator.html.
7. See https://www.ibm.com/ibm/history/ibm100/us/en/icons/tabula tor/.
8. See https://www.ibm.com/ibm/history/ibm100/us/en/icons/tabula tor/. The potential uses of the machines in different industries was reflected in the patents that Hollerith filed for them: "my invention is not limited to such a system [for the census] but may be applied in

effecting compilations of any desired series or system of items representing characteristics of persons, subjects, or objects." Quoted at Geoffrey D. Austrian, HERMAN HOLLERITH: FORGOTTEN GIANT OF INFORMATION PROCESSING (1982) 83. A commercial breakthrough for Hollerith came when managers recognized that the machine vastly improved cost accounting in factories: id. at 200–203; and another when department stores started using it to analyze sales data: id. at 203–205.

9. *The Census of the United States*, 63(9) SCIENTIFIC AMERICAN 132 col. 2 (Aug. 30, 1890).

10. Id. The machine's versatility was recognized from the start. Robert P. Porter, Superintendent of the U.S. Census for 1890, reported to the Secretary of the Interior that the machine allowed "the facts [to] be presented in a greater variety of ways" than heretofore practical: Porter to the Secretary of the Interior, July 3, 1889, as quoted by Austrian at 49. Cf. id. at 64–65, 69. Emphasizing the qualitative change that the Hollerith machine brought about, see Norberg (1990). As to the sheer *speed* of the machine, this was demonstrated in Census Bureau tests in which it handily beat Hollerith's two best contemporary rivals: Austrian at p. 51.

11. Austrian at 124.

12. *Holmes to Pollock* (June 10, 1923), reprinted De Wolfe Howe (ed.) (1942) 118.

13. Holmes's later critic, Lon Fuller, who understood legal realism to fail as a philosophy of law because of its tendency to identify a normative force in facts, picked out the Hollerith machine as an emblem of the realists:

> The facts most relevant to legal study will generally be found to be what may be called moral facts. They lie not in behavior patterns, but in attitudes and conceptions of rightness, in the obscure taboos and hidden reciprocities which permeate business and social relations. They are facts of a type which will not pass readily through a Hollerith statistical sorting machine…

Lon Fuller, *Lecture II*, JULIUS ROSENTHAL LECTURES, NORTHWESTERN UNIVERSITY: THE LAW IN QUEST OF ITSELF (1940) 45, 65.

14. *Dr. Oliver Wendell Holmes at Cambridge University*, 22(551) SCIENTIFIC AMERICAN 8806 col. 2 (July 24, 1886).

15. The Holmes materials in the Harvard Law School Library Digital Suite include a receipt for Holmes's 1927 subscription to the SCIENTIFIC AMERICAN; and it appears from the inventory in his estate that he was clipping articles from the journal in 1922 and receiving it in his library in 1913. See 3:HLS.Libr:7678129 seq. 151; 3:HLS.Libr:8582493 seq. 37; HLS.Libr:8268117 seq. 39.

16. As to which, see Smith Rinehart, *Holmes on Patents*, 98 J. PAT. TRADEMARK OFF. SOC'Y 896 (2016).

17. Consider for example the dispute involving the Census Office itself, Hollerith's former customer: Austrian at 264.
18. Hollerith's Tabulating Machine Company had its headquarters and workshops from 1892 to 1911 at 1054 31st Street, NW: Austrian, pp. 97–99; and see photograph id. between pp. 182 and 183. On first arriving in Washington in December 1902 to begin service as Associate Justice, Holmes and his wife lived at 10 Lafayette Square: Catherine Drinker Brown, YANKEE FROM OLYMPUS: JUSTICE HOLMES AND HIS FAMILY (1945) 353. They later moved to 1720 I Street, which Holmes acquired and refurbished in 1902: Budiansky 286. Senator Henry Cabot Lodge, Jr., writing after Holmes's death to Edward J. Holmes, Holmes's nephew, thought the house on I Street, which Holmes had bequeathed to the government, could be used as a "shrine to [Holmes's] memory" (letter dated June 6, 1939): 3:HLS.Libr:8582488 seq. 53. Edward said the idea would not have pleased Holmes: letter to Lodge dated June 16, 1939, HLS.Libr:8582488 seq. 55. Nothing came of plans to preserve the house. A nondescript office building now occupies the site. The exterior walls of the former warehouse in which the Tabulating Machine Company did its work still exist. IBM placed a plaque there in 1984 to note the connection.
19. Austrian, pp. 60–62.
20. Characteristic was his excitement, expressed in a letter to Lady Pollock, about an outing to see a troupe of jugglers: *Holmes to Lady Pollock* (May 13, 1898), reprinted De Wolfe Howe (ed.) (1942) 87. And he walked to and from work (at least as late as his 70s, which is to say he was still commuting by foot when the 1910 census was counted). See, e.g., Holmes's reference to walking home past the Treasury: Holmes, *Law and the Court, Speech at a Dinner of the Harvard Law School Association of New York* (Feb. 15, 1913), in Posner (ed.) (1992) 145, 148. It also appears that Holmes and his wife at least on one occasion had an outing along the C & O Canal, the waterway on which the Hollerith building is located: Budiansky 281.
21. See example, regarding science and bureaucracy in mid-nineteenth century France, Fox, THE SAVANT AND THE STATE: SCIENCE AND CULTURAL POLITICS IN NINETEENTH-CENTURY FRANCE (2012) 94–137; regarding the systematized survey of its new borders after the partitions of Poland, see Olesko, *The Information Order of the Prussian Frontier, 1772–1806* (Max-Planck-Institut für Wissenschaftsgeschichte, 2019).
22. Hollerith naturally comes up in connection with the growth and intensification of public bureaucracy. See, e.g., BIOGRAPHY OF AN IDEAL: A HISTORY OF THE FEDERAL CIVIL SERVICE (U.S. Civil Service Commission, Office of Public Affairs, 1973) 176. See also Beniger, THE CONTROL REVOLUTION: TECHNOLOGICAL AND ECONOMIC ORIGINS OF THE INFORMATION SOCIETY (1986) 399–400 and *passim*.

23. Not to be confused with eminent historian of modern Europe, Franklin L. Ford (1920–2003).
24. *Holmes to Ford* (Apr. 26, 1907), quoted in Burton (1980) 204.
25. Holmes, *Law and the Court, Speech at a Dinner of the Harvard Law School Association of New York* (Feb. 15, 1913), in Posner (ed.) (1992) 145, 148. Elsewhere, Holmes suggested he was no fan of Wagner: see Budiansky 194–95, 424.
26. At least in any performance of the opera as staged to date.
27. The weaving characters at the start of *Götterdämmerung* were called Norns, Old Norse for Fates. See Richard Wagner, GÖTTERDÄMMERUNG (TWILIGHT OF THE GODS) (Stewart Robb, trans.) (London: Scribner, 1960) 1–2.
28. 12 HARV. L. REV. at 462–63.
29. Posner, 110 HARV. L. REV. 1039, 1040 (1997).
30. Judge Posner quotes the passage here: 110 HARV. L. REV. 1040 n 3; and (1997) 63 BROOK. L. REV. 7, 14–15 (1997).
31. See for example Voegelin, *The New Science of Politics* (originally the 1951 Walgreen Foundation Lectures), republished in MODERNITY WITHOUT RESTRAINT, COLLECTED WORKS, vol. 5 (2000); *The Origins of Scientism*, 15 SOC. RES. 473–476 (1948).
32. See for example Bostrom, *Existential Risks: Analyzing Human Extinction Scenarios and Related Hazards*, 9 J. EVOL. TECH. (2002).
33. For an arch-critic, see Alschuler (1997). An admirer who nevertheless critiques Holmes on this score is Posner: see 110 HARV. L. REV. at 1042 (1997).
34. Gottfried Wilhelm Leibniz, *Dissertatio de Arte Combinaoria* (1666), quoted by Livermore in *Rule by Rules*, chapter in Whalen (ed.) (2019) 3.
35. Livermore (2019).
36. Reprinted in H.L.A. Hart, ESSAYS IN JURISPRUDENCE AND PHILOSOPHY (1983) 22.
37. Hart, *Positivism and the Separation of Law and Morals*, 71 HARV. L. REV. 593, 607 (1958).
38. Adopted April 27, 2016, applicable from May 25, 2018.
39. AB375, Title 1.81.5 (signed into law June 28, 2018, to be operative from Jan. 1, 2020).
40. For a survey, see U.S. Library of Congress, Regulation of Artificial Intelligence in Selected Jurisdictions (January 2019).
41. 71 HARV. L. REV. 593.
42. Lon Fuller was one of the main critics. See Hart's 1957 Holmes Lecture, *Positivism and the Separation of Law and Morals*, 71 HARV. L. REV. 593–629 (1958). See also Sebok (2009).
43. Nicola Lacey, in her A LIFE OF H.L.A. HART: THE NIGHTMARE AND THE NOBLE DREAM (2006), notes the close involvement of Hart with

Bletchley Park during World War II, but she does not note any particular personal ties between Hart and Turing: LACEY 90–93. A personal connection is suggested, in passing, in Gavaghan, DEFENDING THE GENETIC SUPERMARKET: THE LAW AND ETHICS OF SELECTING THE NEXT GENERATION (London: Routledge, 2007) 37. Lacey, who worked closely with Hart's personal papers, is not aware of any closer tie between Hart and Turing apart from the circumstantial one of Hart's role as MI5's ULTRA liaison: correspondence from Professor Lacey to T.D. Grant (Aug. 9, 2019).

44. Finnis notes that Hart "patriotically maintained the mandated secrecy about these activities [for MI5 during the war] down to the end of his life." John Finnis, *H.L.A. Hart: A Twentieth Century Oxford Political Philosopher: Reflections by a Former Student and Colleague*, 54 AM. J. JURISPRUD. 161 (2009).

45. Correspondence from Adam Hart to T.D. Grant (Aug. 15, 2019).

46. Recall Chapter 9, p. 111, n. 17.

Epilogue: Lessons in Two Directions

For an analogy to be worthwhile, it should tell one side or the other—the lawyer or the computer scientist here—something about the other side's discipline that she (i) does not know or understand; and (ii) wishes to know or understand, or should. What have we, the two authors, learnt from this analogy?

A Data Scientist's View

This time last year, I knew nothing about Holmes. I was astonished to find out that the big debate between statistical inference and machine learning, currently being played out in university departments and academic journals, was prefigured by a nineteenth century lawyer. It's impressive enough that Holmes argued for pattern finding from experience rather than logic, well before the birth of modern statistics in the 1920s. It's even more impressive that he took the further leap from scientific rule-inference to machine-learning style prediction.

Holmes, when he opened the door to experience and pattern finding and prediction, seems not to have been troubled by the implications. It seems he had a heady Victorian confidence in science, and he was quite sure that he could perfectly well put together a page of history worth as much as a volume of logic. But the arguments that Holmes let in when he opened that door—arguments about bias, validity, explanation, and so on—are still rumbling, even though legal thinking has had 120 years to deal with them.

© The Editor(s) (if applicable) and The Author(s) 2020
T. D. Grant and D. J. Wischik, *On the path to AI*,
https://doi.org/10.1007/978-3-030-43582-0

Will machine learning still be dealing with these arguments in the year 2140? At the moment it hasn't even caught up with some of the sophisticated ideas that legal thinking has generated so far. And maybe these arguments can never be resolved: maybe it's impossible to hand over responsibility to a formal system, and the burden is on every one of us to learn how to think better with data and experience. Nevertheless, machine learning has a useful brusqueness: ideas that lead to working code are taken forwards, other ideas are sooner or later left by the wayside. It also has exponentially growing datasets and computing power, which give it an ability to step up the abstraction ladder in a way that the law cannot. So I am optimistic that machine learning will lead to intellectual advances, not just better kitten detectors.

Holmes would surely find the present day a most exciting time to be alive.

DJW, August 2019

A Lawyer's View

I was in a seminar in the early 1990s at Yale, taught by the then law school dean Guido Calabresi, where a fellow student suggested a comparison between software and statutory text. I don't recall the detail, except that the comparison was rather elaborate, and Calabresi, impatient with it because it wasn't going anywhere, tried to move the conversation on, but the student persisted—pausing only to say, "No, wait. I'm on a roll." Seeing an opening, the dean affected a dramatic pose; turned to the rest of us; and, over the student who otherwise showed no sign he'd stop, pleaded in tremulous tone: "*You're on a roll?*" The scene sticks with me for its admonitory value: don't try describing law with computer analogies.

In the quarter century since, however, neither law nor computer science has left the other alone. The current debates over explainability, accountability, and transparency of computer outputs point to mutual entanglement. Doing something about decisions reached with machine learning has entered the legislative agenda, and litigators have machine learning and "big data" on their radar. Meanwhile, engineers and investors are looking for ways to make machine learning do ever more impressive things. The rest of us, lawyers included, in turn, scramble to respond to the resultant frictions and figure out, in day to day terms, what machine learning means for us. Could we be missing something basic in the urgency of the moment?

Hearing from people in Cambridge and elsewhere who work on AI and machine learning, I had gotten an intuitive sense that it isn't helpful to talk about computers getting smarter. It didn't sound fine-grained enough to convey what I reckoned an educated layperson ought to know about the topic, nor did it sound quite to the point. Accounts from different specialists over a couple of years added to the picture for me, but I couldn't shake the feeling that there might be something basic missing in even the specialists' appreciation of what they are describing.

To no conscious purpose having anything to do with machine learning, over the holidays in December 2018 I took to re-reading some of Holmes's work, including *The Path of the Law*. A seeming connection roused my curiosity: between Holmes's idea of experience prevailing over logic and machine learning's reliance on data instead of software code. Maybe it was a nice point, but probably not more than that.

Considering Holmes more closely, however, I started to wonder whether there might be something useful in the comparison. The further I looked, the more Holmes's ideas about law seemed to presage problems in machine learning, including, as I came to learn, some that aren't widely known. To take a relatively familiar problem, there is the risk to societal values when a decision-maker is obscure about how he or she (or it) reached a decision. Scholars over the years have noticed in Holmes's work a seeming unconcern about values. Noted at times, but not as often, has been Holmes's concern over how to explain decisions. Holmes resonated as I thought about the current debate over explainability of AI outputs.

Sometimes missed altogether when people read Holmes is the fullness, in its forward-leaning, of the idea that law is prophecy. I came to understand that machine learning isn't just a better way to write software code; it's a way of re-formulating questions to turn them into prediction problems, and, as we've argued, the links to Holmes's idea of prophecy are remarkable. Venturing to see where the analogy to Holmes might go, and testing it with my co-author, I came to appreciate what machine learning does today that's so remarkable—and, also, what it has not yet done.

Holmes's "turn toward induction" is an antecedent to a situation in law that lawyers in the United States call a crisis. A turn toward Holmes, however, might be helpful, whatever one's legal philosophy, for some light it casts on machine learning.

TDG, August 2019

SELECTED BIBLIOGRAPHY

Al-Abdulkarim, Latifa, Katie Atkinson & Trevor Bench-Capon, *Factors, Issues and Values: Revisiting Reasoning with Cases*, International Conference on AI and Law 2015, June 8–12, 2015, San Diego, CA: https://cgi.csc.liv.ac.uk/~tbc/publications/FinalVersionpaper44.pdf.

Alschuler, Albert W., *The Descending Trail: Holmes' Path of the Law One Hundred Years Later*, 49 FLA. L. REV. 353 (1997).

Alschuler, Albert W., LAW WITHOUT VALUES: THE LIFE, WORK, AND LEGACY OF JUSTICE HOLMES (Chicago: University of Chicago Press, 2000).

Arthur, Richard T.W., *The Labyrinth of the Continuum*, in Maria Rosa Antognazza (ed.), OXFORD HANDBOOK OF LEIBNIZ (Oxford: Oxford University Press, 2018) 275.

Artosi, Alberto & Giovanni Sartor, *Leibniz as Jurist* (Antognazza, ed.) (2018) 641.

Ashley, Kevin D., ARTIFICIAL INTELLIGENCE AND LEGAL ANALYTICS: NEW TOOLS FOR LAW PRACTICE IN THE DIGITAL AGE (Cambridge: Cambridge University Press, 2017).

Austrian, Geoffrey D., HERMAN HOLLERITH: FORGOTTEN GIANT OF INFORMATION PROCESSING (New York: Columbia University Press, 1982).

Barocas, Solon & Andrew D. Selbst, *Big Data's Disparate Impact*, 104 CAL. L. REV. 671 (2016).

Barzun, Charles L., *The Positive U-Turn*, 69 STAN. L. REV. 1323 (2017).

Bench-Capon, Trevor et al., *A History of AI and Law in 50 Papers: 25 Years of the International Conference on AI and Law*, 20(3) ART. INTEL. LAW 215 (2012).

© The Editor(s) (if applicable) and The Author(s) 2020   133
T. D. Grant and D. J. Wischik, *On the path to AI*,
https://doi.org/10.1007/978-3-030-43582-0

Beniger, James R., THE CONTROL REVOLUTION: TECHNOLOGICAL AND ECONOMIC ORIGINS OF THE INFORMATION SOCIETY (Cambridge: Harvard University Press, 1986).

Bennett, Thomas B., Barry Friedman, Andrew D. Martin & Susan Navarro Smelcer, *Divide & Concur: Separate Opinions & Legal Change*, 103 CORN. L. REV. 817 (2018).

Benvenisti, Eyal, *Upholding Democracy and the Challenges of New Technology: What Role for the Law of Global Governance?* 29 EJIL 9 (2018).

Berman, Emily, *A Government of Laws and Not of Machines*, 98 B.U. L. REV. 1277 (2018).

Bernstein, Peter L., AGAINST THE GODS: THE REMARKABLE STORY OF RISK (New York: Wiley, 1996).

Bishop, Christopher, PATTERN RECOGNITION AND MACHINE LEARNING (New York: Springer, 2007).

Blackstone, Sir William, COMMENTARIES ON THE LAWS OF ENGLAND: BOOK THE FIRST (Oxford: Clarendon Press, 1765).

Bosmajian, Haig, *Is a Page of History Worth a Volume of Logic?* 38 J. CHURCH STATE 397 (1996).

Bostrom, Nick, *Existential Risks: Analyzing Human Extinction Scenarios and Related Hazards*, 9 J. EVOL. TECH. (2002).

Bostrom, Nick, SUPERINTELLIGENCE: PATHS, DANGERS, STRATEGIES (Oxford: Oxford University Press, 2014).

Breiman, Leo, *Statistical Modeling: The Two Cultures*, 16(3) STATISTICAL SCIENCE 199 (2001).

Brown, Kenneth S. (Gen. Ed.), MCCORMICK ON EVIDENCE (6th edn.) (St Paul: West Publishing, 2006).

Brown, R. Blake & Bruce A. Kimball, *When Holmes Borrowed from Langdell: The 'Ultra Legal' Formalism and Public Policy of Northern Securities (1904)*, 45 AMERICAN JOURNAL OF LEGAL HISTORY 278 (2001).

Brożek, Bartosz & Marek Jakubiec, *On the Legal Responsibility of Autonomous Machines*, 25 ART. INTEL. LAW 293 (2017).

Bryson, Joanna, Mihailis Diamantis & Thomas D. Grant, *Of, for, and by the People: The Legal Lacuna of Synthetic Persons*, 25 ART. INTEL. LAW 273 (2017).

Budiansky, Stephen, OLIVER WENDELL HOLMES: A LIFE IN WAR, LAW, AND IDEAS (New York: W.W. Norton, 2019).

Burri, Thomas, *International Law and Artificial Intelligence*, 60 GYIL 91 (2017).

Burton, David H., *The Curious Correspondence of Justice Oliver Wendell Holmes and Franklin Ford*, 53(2) N. Engl. Q. 196 (1980).

Cardozo, Benjamin, THE NATURE OF THE JUDICIAL PROCESS (New Haven: Yale University Press, 1921).

Casey, Bryan, *The Next Chapter in the GDPR's "Right to Explanation" Debate and What It Means for Algorithms in Enterprise*, EUROPEAN UNION LAW WORKING PAPERS, No. 29 (2018).

Chessman, Christian, *Note. A Source of Error: Computer Code, Criminal Defendants, and the Constitution*, 105 CAL. L. REV. 179 (2017).

Chouldechova, Alexandra & Aaron Roth, *The Frontiers of Fairness in Machine Learning*, sec. 3.3. p. 6 (Oct. 20, 2018): https://arxiv.org/pdf/1810.08810.pdf.

Clarke, Arthur C., *Hazards of Prophecy: The Failure of Imagination*, in PROFILES OF THE FUTURE: AN INQUIRY INTO THE LIMITS OF THE POSSIBLE (1962, rev. 1973) 21.

Cohen, Felix S., *The Holmes-Cohen Correspondence*, 9 J. HIST. IDEAS 3 (1948).

Cook, Nancy, *Law as Science: Revisiting Langdell's Paradigm in the 21st Century*, 88 N.D. L. REV. 21 (2012).

Daston, Lorraine, CLASSICAL PROBABILITY IN THE ENLIGHTENMENT (Princeton: Princeton University Press, 1988).

Dawid, Philip, *Probability and Statistics in the Law*, Research Report 243, Department of Statistical Science, University College London (May 2004).

De Wolfe Howe, Mark (ed.), HOLMES-POLLOCK LETTERS: THE CORRESPONDENCE OF MR. JUSTICE HOLMES AND SIR FREDERICK POLLOCK 1874–1932 (Cambridge: Harvard University Press, 1942).

De Wolfe Howe, Mark (ed.), THE OCCASIONAL SPEECHES OF JUSTICE OLIVER WENDELL HOLMES (Cambridge: Harvard University Press, 1962).

Deng, J., W. Dong, R. Socher, L.-J. Li, K. Li & L. Fei-Fei, *ImageNet: A Large-Scale Hierarchical Image Database*, CVPR (2009).

Dijkstra, Edsger W., *How Do We Tell Truths That Might Hurt?* in Edsger W. Dijkstra, SELECTED WRITINGS ON COMPUTING: A PERSONAL PERSPECTIVE (1982) 129 (original text dated June 18, 1975).

Donoho, David, *50 Years of Data Science* (Sept. 18, 2015), from a Presentation at the Tukey Centennial Workshop, Princeton, NJ.

Dwork, Cynthia, et al., *Fairness Through Awareness*, ITCS CONF. PROC. (3RD) (2012).

Eber, Michael L., *Comment, When the Dissent Creates the Law: Cross-Cutting Majorities and the Prediction Model of Precedent*, 58 EMORY L.J. 207 (2008).

Efron, Bradley & Trevor Hastie, COMPUTER AGE STATISTICAL INFERENCE: ALGORITHMS, EVIDENCE, AND DATA SCIENCE (Cambridge: Cambridge University Press, 2016).

European Parliament, *Civil Law Rules on Robotics Resolution* (Feb. 16, 2017): P8_TA(2017)0051.

Fetzer, James H. (ed.), THE PHILOSOPHY OF CARL G. HEMPEL: STUDIES IN SCIENCE, EXPLANATION, AND RATIONALITY (Oxford: Oxford University Press, 2001).

Fienberg, Stephen E. & Joseph B. Kadane, *The Presentation of Bayesian Statistical Analyses in Legal Proceedings*, THE STATISTICIAN 32 (1983).

Finnis, John, *H.L.A. Hart: A Twentieth Century Oxford Political Philosopher: Reflections by a Former Student and Colleague*, 54 AM. J. JURISPRUD. 161 (2009).

Franklin, James, THE SCIENCE OF CONJECTURE: EVIDENCE AND PROBABILITY BEFORE PASCAL (Baltimore: Johns Hopkins University Press, 2001).

Franklin, James, *Pre-history of Probability*, in Alan Hájak & Christopher Hitchcock (eds.), OXFORD HANDBOOK OF PROBABILITY AND PHILOSOPHY (Oxford University Press, 2016) 33.

Fuller, Lon, *Lecture II*, in JULIUS ROSENTHAL LECTURES, NORTHWESTERN UNIVERSITY: THE LAW IN QUEST OF ITSELF (Chicago: Foundation Press, 1940).

Gillespie, Tarleton, *The Relevance of Algorithms*, in Gillespie et al. (eds.), MEDIA TECHNOLOGIES: ESSAYS ON COMMUNICATION, MATERIALITY, AND SOCIETY 167 (Cambridge: MIT Press, 2014).

Gordon, Robert W., *Introduction*, in Gordon (ed.), THE LEGACY OF OLIVER WENDELL HOLMES JR. (Stanford: Stanford University Press, 1992).

Grey, Thomas C., *Langdell's Orthodoxy*, 45 U. PITT. L. REV. 1 (1983).

Hacking, Ian, THE EMERGENCE OF PROBABILITY (Cambridge University Press, 1975).

Halpern, Mark, *The Trouble with the Turing Test*, THE NEW ATLANTIS (Winter 2006): https://www.thenewatlantis.com/publications/the-trouble-with-the-turing-test.

Hart, H.L.A., *Positivism and the Separation of Law and Morals*, 71 HARV. L. REV. 593 (1958).

Hart, H.L.A., ESSAYS IN JURISPRUDENCE AND PHILOSOPHY (Oxford: Oxford University Press, 1983).

Hastie, Trevor, Robert Tibshirani, & Jerome H. Friedman, THE ELEMENTS OF STATISTICAL LEARNING (2nd edn.) (New York: Springer, 2009).

Hempel, Carl G., ASPECTS OF SCIENTIFIC EXPLANATION AND OTHER ESSAYS IN PHILOSOPHY OF SCIENCE (New York: Free Press, 1965).

Hempel, Carl G., PHILOSOPHY OF NATURAL SCIENCE (Upper Saddle River, NJ: Prentice Hall, 1966).

Hertza, Vlad A., *Fighting Unfair Classifications in Credit Reporting: Should the United States Adopt GDPR-Inspired Rights in Regulating Consumer Credit?* 93 N.Y.U. L. REV. 1707 (2018).

Holmes, Jr., Oliver Wendell, *The Theory of Torts*, 7 AM. LAW REV. 652 (July 1873).

Holmes, Jr., Oliver Wendell, *Book Notice Reviewing a Selection of Cases on the Law of Contracts, with a Summary of the Topics Covered by the Cases, By C.C. Langdell*, 14 AM. LAW REV. 233 (1880).

Holmes, Jr., Oliver Wendell, THE COMMON LAW (1881).

Holmes, Jr., Oliver Wendell, *The Path of the Law*, 10 HARV. L. REV. 457 (1896–1897).

Holmes, Jr., Oliver Wendell, *Law in Science and Science in Law*, 12 HARV. L. REV. 443 (1898–1899).

Holmes, Jr., Oliver Wendell, *The Theory of Legal Interpretation*, 12 HARV. L. REV. 417 (1898–1899).

Holmes, Jr., Oliver Wendell, *Codes, and the Arrangement of the Law*, 5 AM. LAW REV. (October 1870) reprinted in Frederic R. Kellogg, THE FORMATIVE ESSAYS OF JUSTICE HOLMES (1984).

Holmes, Jr., Oliver Wendell & John Jingxiong Wu, JUSTICE HOLMES TO DOCTOR WU: AN INTIMATE CORRESPONDENCE, 1921–1932 (New York: Central Book Company, 1947).

Hossain, MD. Zakir, Ferdous Sohel, Mohd Fairuz Shiratuddin & Hamid Laga, *A Comprehensive Survey of Deep Learning for Image Captioning*, 51(6) ACM CSUR, Article 118 (Feb. 2019), 36 pages: https://doi.org/10.1145/3295748.

House of Lords (UK) Select Committee on Artificial Intelligence, Report (Apr. 16, 2018).

Huhn, Wilson, *The Stages of Legal Reasoning: Formalism, Analogy, and Realism*, 48 VILL. L. REV. 305 (2003).

Immerman, Neil, *Computability and Complexity*, in Edward N. Zalta (ed.), THE STANFORD ENCYCLOPEDIA OF PHILOSOPHY (Winter 2018 edition): https://plato.stanford.edu/archives/win2018/entries/computability/.

Jackson, Vicki C., *Thayer, Holmes, Brandeis: Conceptions of Judicial Review, Factfinding, and Proportionality*, 130 HARV. L. REV. 2348 (2017).

Kamishima, Toshihiro, Shotara Akaho & Jun Sakuma, *Fairness-Aware Learning Through Regularization Approach*, 2011 11th IEEE International Conference on Data Mining Workshops.

Kellogg, Frederic R., THE FORMATIVE ESSAYS OF JUSTICE HOLMES: THE MAKING OF AN AMERICAN LEGAL PHILOSOPHY (Westport: Praeger, 1984).

Kellogg, Frederic R., OLIVER WENDELL HOLMES JR. AND LEGAL LOGIC (Chicago: University of Chicago Press, 2018).

Kimball, Bruce A., INCEPTION OF MODERN PROFESSIONAL EDUCATION: C.C. LANGDELL, 1826–1906 (Chapel Hill: University of North Carolina Press, 2009).

Kozinski, Alex, *What I Ate for Breakfast and Other Mysteries of Judicial Decision Making*, 26 LOY. L.A. L. REV. 993 (1993).

Krizhevksy, Alex, Ilya Sutskever & Geofrey E. Hinton, *ImageNet Classification with Deep Convolutional Neural Networks*, 60(6) COMMS. ACM 84 (2017).

Kroll, Joshua A., *The Fallacy of Inscrutability*, PHIL. TRANS. R. SOC. A 376 (2018).

Kroll, Joshua A., Joanna Huey, Solon Barocas, Edward W. Felten, Joel R. Reidenberg, David G. Robinson & Harlan Yu, *Accountable Algorithms*, 165 U. PA. L. REV. 633 (2017).

Kuhn, Thomas S., *The Structure of Scientific Revolutions: II. The Route to Normal Science*, 2(2) INT. ENCYCL. UNIFIED SCI. 17 (1962).

Lacey, Nicola, A LIFE OF H.L.A. HART: THE NIGHTMARE AND THE NOBLE DREAM (Oxford: Oxford University Press, 2006).

Landes, David S., REVOLUTION IN TIME: CLOCKS AND THE MAKING OF THE MODERN WORLD (Cambridge: Harvard University Press, 1983).

Law Library of Congress (U.S.), Global Legal Research Directorate, *Regulations of Artificial Intelligence in Selected Jurisdictions* (January 2019).

Leibniz, Gottfried Wilhelm, *Dissertatio de Arte Combinaoria* (1666).

Lemaréchal, Claude, *Cauchy and the Gradient Method*, DOCUMENTAL MATH. 251 (2012).

Leonard, David P. (ed.), THE NEW WIGMORE: A TREATISE ON EVIDENCE. SELECTED RULES OF LIMITED ADMISSIBILITY (New York: Aspen Law & Business, 2010).

Lessig, Lawrence, CODE AND OTHER LAWS OF CYBERSPACE (New York: Basic Books, 1999).

Levi, Edward H., AN INTRODUCTION TO LEGAL REASONING (Chicago: University of Chicago Press, 1949).

Livermore, Michael A., *Rule by Rules*, Chapter in Ryan Whalen (ed.), COMPUTATIONAL LEGAL STUDIES: THE PROMISE AND CHALLENGE OF DATA-DRIVEN LEGAL RESEARCH (2019) 2.

Livermore, Michael A. & Daniel N. Rockmore (eds.), LAW AS DATA: COMPUTATION, TEXT, & THE FUTURE OF LEGAL ANALYSIS (Sante Fe: Santa Fe Institute Press, 2019).

Livermore, Michael A., Faraz Dadgostari, Mauricio Guim, Peter Beling & Daniel Rockmore, *Law Search as Prediction* (Nov. 5, 2018), Virginia Public Law and Legal Theory Research Paper No. 2018-61: https://ssrn.com/abstract=3278398.

Madras, David, Elliot Creager, Toniann Pitassi & Richard Zemel, *Learning Adversarially Fair and Transferable Representations* (last revised Oct. 22, 2018): arXiv: 1802.06309 [cs.LG].

Ministry of Industry and Information Technology (People's Republic of China), *Three-Year Action Plan for Promoting Development of a New Generation Artificial Intelligence Industry* (2018–2020) (published Dec. 14, 2017).

Moskowitz, David H., *The Prediction Theory of Law*, 39 TEMP. L.Q. 413, 413 (1965–66).

Murphy, Kevin P., MACHINE LEARNING: A PROBABILISTIC PERSPECTIVE (Cambridge: MIT Press, 2012).

National Science and Technology Council (U.S.), *The National Artificial Intelligence Research Development Strategic Plan* (Oct. 2016).

Norberg, Arthur L., *High-Technology Calculation in the Early 20th Century: Punched Card Machinery in Business and Government*, 31(4) TECHNOLOGY AND CULTURE 753–779 (1990).

Novick, Sheldon M., THE COLLECTED WORKS OF JUSTICE HOLMES: COMPLETE PUBLIC WRITINGS AND SELECTED JUDICIAL OPINIONS OF OLIVER WENDELL HOLMES (with a foreword by Erwin N. Griswold) (Chicago: University of Chicago Press, 1995).

Obermeyer, Ziad & Ezekiel J. Emanuel, *Predicting the Future—Big Data, Machine Learning, and Clinical Medicine*, 375(13) NEJM 1216 (Sept. 29, 2016).

Pascal, Blaise, PENSÉES, Part II, W.F. Trotter (trans.) (London: Dent, 1908).

Pasquale, Frank, THE BLACK BOX SOCIETY: THE SECRET ALGORITHMS THAT CONTROL MONEY AND INFORMATION (Cambridge: Harvard University Press, 2015).

Pasquale, Frank, *Bittersweet Mysteries of Machine Learning (A Provocation)*, LONDON SCH. ECON. POL. SCI.: MEDIA POL'Y PROJECT BLOG (Feb. 5, 2016).

Pitler, Robert M., *"The Fruit of the Poisonous Tree" Revisited and Shepardized*, 56 CAL. L. REV. 579 (1968).

Pohlman, H.L., JUSTICE OLIVER WENDELL HOLMES & UTILITARIAN JURISPRUDENCE (Cambridge: Harvard University Press, 1984).

Posner, Richard A., THE PROBLEMS OF JURISPRUDENCE (Cambridge: Harvard University Press, 1990).

Posner, Richard A., THE ESSENTIAL HOLMES: SELECTIONS FROM THE LETTERS, SPEECHES, JUDICIAL OPINIONS, AND OTHER WRITINGS OF OLIVER WENDELL HOLMES, JR. (Chicago: University of Chicago Press, 1992).

Posner, Richard A., *The Path Away from the Law*, 110 HARV. L. REV. 1039 (1997).

Posner, Richard, *Legal Scholarship Today*, 115(5) HARV. L. REV. 1314 (2002).

Rabban, David M., *Holmes the Historian*, in Rabban (ed.), LAW'S HISTORY: AMERICAN LEGAL THOUGHT AND THE TRANSATLANTIC TURN TO HISTORY (Cambridge: Cambridge University Press, 2013).

Randell, Brian, *The History of Digital Computers*, 12(11–12) IMA BULL. 335 (1976).

Rosenblatt, Frank, *The Perceptron: A Probabilistic Model for Information Storage and Organization in the Brain*, 65(6) PSYCHOLOGICAL REVIEW 386 (1958).

Roth, Andrea, *Machine Testimony*, 126 YALE L. J. 1972 (2017).

Russakovsky, Olga*, Jia Deng*, Hao Su, Jonathan Krause, Sanjeev Satheesh, Sean Ma, Zhiheng Huang, Andrej Karpathy, Aditya Khosla, Michael Bernstein, Alexander C. Berg & Li Fei-Fei. (* = equal contribution) *ImageNet Large Scale Visual Recognition Challenge*. IJCV, 2015.

Russell, Stuart J. & Peter Norvig, ARTIFICIAL INTELLIGENCE: A MODERN APPROACH (3rd edn.) (Harlow: Pearson Education, 2016).

Saxe et al. *On the Information Bottleneck Theory of Deep Learning* (ICLR, 2018).

Schauer, Frederick & Barbara A. Spellman, *Analogy, Expertise, and Experience*, 84 U. CHI. L. REV. 249 (2017).

Schneider, G. Michael & Judith Gersting, INVITATION TO COMPUTER SCIENCE (New York: West Publishing Company, 1995).

Sebok, Anthony J., *Comment on 'Law as a Means'* and other essays in Peter Cane (ed.), THE HART/FULLER DEBATE AT FIFTY (Oxford: Hart Publishing, 2009).

Sedgewick, Robert & Kevin Wayne, ALGORITHMS (4th ed.) (New York: Addison-Wesley, 2011).

Shmueli, Galit, *To Explain or to Predict?* 25(3) STATISTICAL SCIENCE 289 (2010).

Shwartz-Ziv, R. & N. Tishby, *Opening the Black Box of Deep Neural Networks via Information* (arXiv, 2017).

Silver, David, Aja Huang, Chris J. Maddison, et al., *Mastering the Game of Go with Deep Neural Networks and Tree Search*. 529 NATURE, 484–489 (2016).

Silver, David, J. Schrittwieser, K. Simonyan, et al. *Mastering the Game of Go Without Human Knowledge*, 550 NATURE, 354–359 (2017). https://doi.org/10.1038/nature24270.

Smith Rinehart, Amelia, *Holmes on Patents: Or How I Learned to Stop Worrying and Love Patent Law*, 98 J. PAT. TRADEMARK OFF. SOC'Y 896 (2016).

Suk Gersen, Jeannie, *The Socratic Method in the Age of Trauma*, 130 HARV. L. REV. 2320 (2017).

Swade, Doron, DIFFERENCE ENGINE: CHARLES BABBAGE AND THE QUEST TO BUILD THE FIRST COMPUTER (New York: Viking Penguin, 2001).

The Census of the United States, 63(9) SCIENTIFIC AMERICAN 132 col. 2 (Aug. 30, 1890).

Turing, Alan M., *On Computable Numbers, with an Application to the Entscheidungsproblem*, 2(42) LMS PROC. (1936).

Turing, Alan M., *Computing Machinery and Intelligence*, 59 MIND 433 (1950).

Tyree, Alan L., EXPERT SYSTEMS IN LAW (New York: Prentice Hall, 1989).

Unger, Roberto M., LAW IN MODERN SOCIETY: TOWARD A CRITICISM OF SOCIAL THEORY (New York: The Free Press, 1976).

Urbaniak, Rafal, *Narration in Judiciary Fact-Finding: A Probabilistic Explication*, 26 ART. INTEL. LAW 345 (2018).

Voegelin, Eric, *The Origins of Scientism*, 15 SOCIAL RESEARCH 473 (1948).

Voegelin, Eric, *The New Science of Politics* (originally the 1951 Walgreen Foundation Lectures), republished in MODERNITY WITHOUT RESTRAINT, COLLECTED WORKS, vol. 5 (2000).

Wagner, Ben, Rapporteur, Committee of Experts on Internet Intermediaries (MSI-NET, Council of Europe), *Study on the Human Rights Dimensions of Algorithms* (2017).

Wendel, W. Bradley, *Explanation in Legal Scholarship: The Inferential Structure of Doctrinal Legal Analysis*, 96 CORN. L. REV. 1035 (2011).

Yang, Kaiyu, Klint Qinami, Li Fei-Fei, Jia Deng, Olga Russakovsky, *Towards Fairer Datasets: Filtering and Balancing the Distribution of the People Subtree in the ImageNet Hierarchy*, research post from Sept. 17, 2019, http://image-net.org/update-oct-13-2019 (retrieved Jan. 12, 2020).

Zalta, Edward N. (ed.), THE STANFORD ENCYCLOPEDIA OF PHILOSOPHY (Summer 2019 edition).

Žliobaitė, Indrė & Bart Custers, *Using Sensitive Personal Data May Be Necessary for Avoiding Discrimination in Data-Driven Decision Models*, 24 ART. INTEL. LAW 183 (2016).

Zobel, Hiller B., *Oliver Wendell Holmes, Jr., Trial Judge*, 36 BOSTON BAR J. 25 (March–April 1992).

Zobel, Hiller B., *Justice Oliver Wendell Holmes in the Trial Court*, 8 MASSACHUSETTS LEGAL HISTORY 35 (2002).

INDEX

© The Editor(s) (if applicable) and The Author(s) 2020
T. D. Grant and D. J. Wischik, *On the path to AI*,
https://doi.org/10.1007/978-3-030-43582-0